Making Sense of Sensemaking

Making Sense of Sensemaking

Designing Authentic K–12 STEM Learning Experiences

TJ McKenna

TEACHERS COLLEGE PRESS
TEACHERS COLLEGE | COLUMBIA UNIVERSITY
NEW YORK AND LONDON
1234 AMSTERDAM AVENUE
NEW YORK, NY 10027

nsta Press®
National Science Teaching Association
405 E LABURNUM AVENUE
STE 3
RICHMOND, VA 23222

Published simultaneously by Teachers College Press, 1234 Amsterdam Avenue, New York, NY 10027 and National Science Teaching Association, 405 E Laburnum Avenue Ste 3, Richmond, VA 23222.

Front cover design by Peter Donahue. Illustration by Kittisak Taramas/iStock by Getty Images. Figures at table by Shamim Art / Shutterstock.

Library of Congress Cataloging-in-Publication Data

Names: McKenna, T. J. author
Title: Making sense of sensemaking : designing authentic K-12 STEM learning experiences / TJ McKenna.
Description: New York, NY : Teachers College Press, [2025] | Includes bibliographical references and index. | Summary: "This guide describes what sensemaking is, why it is important, and how to design dynamic learning experiences that are student-centered"— Provided by publisher.
Identifiers: LCCN 2024058115 (print) | LCCN 2024058116 (ebook) | ISBN 9780807786871 hardcover | ISBN 9780807786864 paperback | ISBN 9780807782965 ebook
Subjects: LCSH: Student-centered learning | Science—Study and teaching | Technology—Study and teaching | Mathematics—Study and teaching
Classification: LCC LB1027.23 .M395 2025 (print) | LCC LB1027.23 (ebook) | DDC 371.39/4—dc23/eng/20250220
LC record available at https://lccn.loc.gov/2024058115
LC ebook record available at https://lccn.loc.gov/2024058116

ISBN 978-0-8077-8686-4 (paper)
ISBN 978-0-8077-8687-1 (hardcover)
ISBN 978-0-8077-8296-5 (ebook)

Printed on acid-free paper
Manufactured in the United States of America

To everyone who has helped me along my own journey of sensemaking, especially my wife, Stephanie, who is always my first stop on the path to seeing if my outlandish ideas make any sense.

Contents

Introduction

In a 5th-grade classroom, students are trying to explain what will happen to two volleyballs that initially weigh the same on a balance, after one has had 10 more pumps added from a hand air pump. One student shares that they think it will be lighter because air makes things lighter—like a balloon. Another student says that air must weigh something, and adding air will make it heavier. A third student says the balance will not change because air is not heavy, and the two volleyballs will weigh the same. A version of this experiment is a common one in many K–12 classrooms.

Now picture that this same experiment is happening in a room full of close to 100 educational leaders (school principals, STEM coaches, science teachers and leaders, etc.). The same three possibilities emerge related to the effect of additional air pumped into one of the two volleyballs—with some stating it will be lighter, some thinking it will weigh the same, and some believing it will be heavier. Why do these students and teacher leaders both share the same lines of reasoning even though they are at such different points in their education? The answer lies in a topic that is challenging to describe, yet increasingly discussed in science education—sensemaking. Clearly, not all of these students or educational leaders are correct in their predictions. But why is this an important discussion to have? That is, why is talking about wrong answers important to figuring out what actually happens in the real world? In this case and all others, it is because sensemaking lies at the heart of doing real science both in the classroom and beyond.

A key moment in my own sensemaking began when I encountered the phenomenon of lying squirrels. Strange, right? I have been involved in science my whole life, really wanting to be a part of anything science-related—informal or formal—but it was a chance encounter with an animal behavior researcher that changed my life. In hindsight, it should not be surprising because at 5 years of age, I told my mother that I wanted to study animal behavior. I remember her laughing and saying, "That's not a job!" Instead, she advised that I needed to be a teacher like her.

Jumping ahead to my undergraduate years when I was studying to be a high school biology teacher, I was walking across campus and noticed a woman with a notebook following some squirrels around and eagerly taking notes. Gathering the confidence to approach her, I asked, "What are you doing?" She shared that squirrels do this pretty crazy thing where they seem to know when potential food-stealers are around and actively *pretend* to bury nuts so they can deceive the would-be thieves and successfully store their food for retrieval later. Intrigued, I asked why they do this, and the professor's response was, "We have some ideas, but honestly we do not know. We're out here trying to figure that out."

At that moment, it clicked in my mind that all areas of science could be similar to this—figuring out things we do not know. Scientists are doing this right now—as I am writing these words, as you are reading them—and the need to figure things out about our world will continue to happen around the globe as people naturally apply their own sense of wonder to phenomena in the natural world. I was immediately hooked by this idea, and I joined the squirrel researcher, Dr. Sylvia Halkin, to try to figure out why the squirrels maintained the habit of pretending to bury their food when potential food robbers were in the area. In the years that followed, we studied this topic, and I was taken by my conversations with people from all walks of life who were also excited to talk about what they observed informally in watching squirrels. Everyone had a story, a hunch, or an example that led them to claim something about squirrels. Their notions ranged from simple backyard observations to hilarious accounts of different squirrels going to insane lengths to figure out so-called "squirrel-proof" bird feeders. Yet many of these people did not see themselves as scientists or even as individuals who liked science. Rather, they just saw themselves as individuals who were *interested* in the world around them.

While our research on squirrels was eventually published in the *Journal of Animal Behaviour* (Steele et al., 2008), I was drawn to helping others see that their own interests can be at the heart of doing science. This led me to be the first staff scientist hired at the Connecticut Science Center, where I spent almost 10 years designing programs, exhibits, and experiences that presented the scientific process to the masses. I co-hosted a weekly TV segment on our local CBS affiliate titled "Science Sunday," during which I aimed to be a resource for teachers and parents who did not think their kids were "into science." In this endeavor, I aimed to be the opposite of a Bill Nye type of on-air scientist who gave away all the answers, rather than introduce viewers to interesting and engaging phenomena that would leave them with new questions and experiments to pursue on their own. As a small indication of my success in advancing scientific wonder on TV, when my wife and I would go to the grocery store on a Sunday morning (often right after "Science Sunday" aired), groups of families would come up to me, saying, "Hey, you're the science guy! We're here to buy the things you talked about so we can experiment on our own!" To me, this was a sign that the phenomenon I shared was interesting enough to motivate this family to do their own sensemaking. If I were to have just told them the answer during the segment (as Bill Nye often did), then I doubt they would have given it a second thought. If done right, they would be motivated to experiment on their own. This is an important distinction because figuring something out for yourself is actually when the knowledge sticks.

If you have ever thought about something you do not understand and then immediately Google it, how does it feel when you find out the answer? To me, it always feels unrewarding. *Yup, cool—no need for sensemaking.* But if you actually take the time to discuss something in detail with someone—like the fact that many clouds weigh over a *billion pounds* and yet just float around effortlessly in the sky—you will often be surprised by the twists and turns involved in that conversation. Usually it starts with someone saying, "That can't be real." After assuring the person that it is real, people often recall a bit of knowledge they forgot they even knew, but all of a sudden seems relevant to this topic. Other times, someone brings up something they are reminded of—a connection they are making that might point the conversation in a new direction. Depending on the relationship between those involved in such a conversation, you might even feel comfortable saying something like, "Yeah, I really never understood how buoyancy works in the water versus in the air. Is floating in air the same as floating in water?" This is what a sensemaking discussion can sound like—messy, using both common and technical language; sometimes disagreeing, often with a lot of clarifying; and usually including at least one *huh*! or *wait*!*; what, that's crazy*!*;* or *that can't be real, true*!

These kinds of sensemaking discussions are very similar to what doing science looks and feels like. When I later presented our squirrel research to a room full of behavioral researchers, I could not wait to get to the end and hear the sound of applause from my audience. In reality, when I got to my last presentation slide, there was no applause. Instead, what looked like every hand in the room shot up with a question or comment. Questions were about our sample size, our methodology, how we analyzed the data, and how we reached our conclusions. It was not until after I had answered each and every question that this group of scientists gave me a nod of approval and began to clap. I realized that science does not end with doing the experiments, but instead requires you to go public with your initial conjectures. It takes explaining the background research you did to figure out what was already known about this topic, how you chose to study the topic, how you analyzed your data, what you ultimately concluded, and what questions still remain or what you want to study next. Simply put, it is essential to explain your sensemaking journey. Where did it start? What did you do? And where did you end up?

This book is about that journey. How do we start sensemaking in K–12 classrooms? What does a shift to sensemaking mean for the roles of teachers and students? How can educational leaders and professional learning providers better support sensemaking? And how does teaching this way make science education more equitable for all learners? While this book is focused on sensemaking, the *Framework for K–12 Science Education* (National Research Council [NCR], 2013) describes a broad view of a new vision for science education, laying out an approach for raising the bar to reach the Framework's ambitious goals. In what follows, I give a brief overview of this vision before digging into sensemaking in more detail.

A NEW VISION FOR SCIENCE EDUCATION

A Framework for K–12 Science Education (NRC, 2013; hereafter referred to as the Framework) and the Next Generation Science Standards (NGSS; NGSS Lead States, 2013) build on 30 years of research in cognitive science, science education, science and technology studies, and education in general, together presenting a vision of rich science education for *all* students. The reforms called for in the Framework and the NGSS, while not entirely new, still represent a major shift from earlier standards and necessitate substantial changes in teachers' practices. Realizing this vision will ultimately require changing our approach to science teaching and learning.

Many describe this change as a shift from "learning about" science to "figuring out" science. The Framework calls this shift three-dimensional science teaching and learning, whereby the science and engineering practices (SEPs) are used to figure out disciplinary core ideas (DCIs), along with the crosscutting concepts (CCCs) functioning as a set of lenses with which to make connections. Sensemaking arises from careful integration of these three dimensions. This approach is student-centered, practice-based, and focused on real-world problems and scenarios. The committee that wrote the Framework (upon which the NGSS was based) carefully considered which science ideas and practices are core to students becoming scientifically literate citizens who can engage with scientific issues in an informed way. The CCCs provide a set of ways to look at these core ideas, sometimes from a new perspective. For example, consider elementary students who are studying waves (light and sound) and trying to model how waves can make something move (vibrate). Rather than having the teacher define waves and teach students about how wavelength and amplitude are involved, students themselves are exploring a room full of speakers and objects that make sounds with the goal of finding patterns (a CCC) that they can use to build a general model of wave behavior. In a second example, students are using a crosscutting concept and the science practices to identify similarities and differences in the patterns they are observing in order to describe what they observe when these patterns change (i.e., the amplitude increases and changes the movement of an object). In this scenario, students are engaging in a carefully designed learning experience in which sensemaking is at the center, and they are motivated to figure out what is happening. If you were a fly on the wall when sensemaking like this is occurring, you might see students talking excitedly about what they do not understand. Thus, examples of science and engineering practices include SEP 1: Asking questions, students disagreeing about elements that they believe are important; SEP 7: Engaging in argument from evidence, or even students digging into text or media to gather more ideas; and SEP 8: Obtaining, evaluating, and communicating information that they can use to make sense of the patterns in these waves. (A complete list of the SEPs can be found in Appendix F of the NGSS: https://www.nextgenscience.org/sites/default/files/Appendix%20F%20%20Science%20and%20Engineering%20Practices%20in%20the%20NGSS%20-%20FINAL%20060513.pdf.)

But how do we know if what we are seeing is sensemaking or confusion? How can we differentiate between academically productive sensemaking and a never-ending search for an answer that always seems out of reach? How do we support our students in exploring not just *what* happens in the world, but also *how* and *why* these things happen, that is, the mechanisms behind phenomena (Braaten & Windschitl, 2011)? And how can we design lessons and units that meaningfully engage students in sensemaking when a quick Google search or conversation with ChatGPT seems to give away too much? In order to make progress on these questions, we must first start with a definition of sensemaking.

WHAT IS SENSEMAKING, AND WHY IS IT IMPORTANT?

Sensemaking is a cognitive and sociocultural process through which individuals actively interpret information or experiences to develop coherent and meaningful understandings of complex and ambiguous phenomena. In essence, sensemaking is the cognitive mechanism we use to comprehend confusing or unclear phenomena and solve problems. This multifaceted process encompasses perceptual observation, interpretation, retrospection, social interaction, and decision-making—all underscoring its iterative and nonlinear nature. Sensemaking resembles assembling a mental puzzle whereby we gather information, attempt to make sense of it by attributing meaning, reflect on prior knowledge, engage with others to exchange ideas, and decide on actions based on our enhanced understanding. Sensemaking necessitates both individual cognition and social construction, highlighting the interplay involving perception, interpretation, and collaborative efforts in creating meaning amid uncertainty and complexity.

While there have been few attempts at describing what sensemaking is across the science education literature, the following definition by Odden and Russ (2019) captures a number of the essential components:

> Sensemaking is a dynamic process of building or revising an explanation in order to "figure something out"—to ascertain the mechanism underlying a phenomenon in order to resolve a gap or inconsistency in one's understanding. One builds this explanation out of a mix of everyday knowledge and formal knowledge by iteratively proposing and connecting up different ideas on the subject. One also simultaneously checks that those connections and ideas are coherent, both with one another and with other ideas in one's knowledge system. (pp. 192–193)

In this definition, science is the process of constructing and critiquing explanations (Ford, 2012). While this may seem oversimplified, all sensemaking involves building explanations by connecting evidence to claims and checking to see that any new evidence either supports or refutes our current explanations. As these explanations grow to include new evidence, you must also check to make sure that all the elements are coherent with one another. In this way, critiquing your own work and that of others is crucial to ensuring that your explanations are aligned with everything you know up to this point.

THE ROLES OF PHENOMENA AND PROBLEMS

One of the first major takeaways for those who are new to the NGSS often revolves around the role of phenomena and problems. I had the chance to see this firsthand as I was leading a combined statewide project and research study on the role of science professional development (PD) facilitators and spent multiple years traveling around the state of Connecticut working with groups who were new to the NGSS. This ultimately led to me founding the website NGSS Phenomena during a lunch break of one of these science-focused professional learning sessions that happens as a part of the Next Generation Science eXemplar (NGSX) System.

I have been involved with the NGSX for many years, starting as a participant and moving up into the role of a national facilitator and lead designer for new pathways. NGSX is a professional learning system designed to support K–12 teachers and science teacher leaders in learning about, engaging in, and bringing the practices articulated in the NGSS to the classroom. In 2013, Moon, Michaels, and Reiser began NGSX with support from the National Science Foundation as a response to the pedagogical changes called for by the NGSS (Moon et al., 2012; Reiser, 2013). NGSX took a systemic approach to change that included the NGSS standards, professional development for teachers, work with state policymakers, and development and use of a statewide network of PD facilitators. I am now a co-director of NGSX and director of the Boston University (BU) Wheelock Center for STEM Professional Learning at Scale, the institutional home of NGSX. For the more than 16,000 educators who have completed one of the multiday pathways, the number one takeaway is always related to phenomena. During a pathway session titled "Becoming a Next Gen Science Teacher," educators viewed a video of a tanker car that collapsed after being steam-cleaned. They then viewed an experiment with a 2-liter bottle that did something strikingly similar when filled with a small amount of boiling water and then capped. This is the kind of phenomenon that participants worked to explain over the course of 4 and a half days dedicated to that PD pathway. Early on in the PD pathway, participants found that their explanations of these phenomena left gaps in their own understanding. They shared these concerns in groups of elementary, middle, and high school teachers. When pressed, they often resorted to using vocabulary like *negative pressure* and *suction* to explain why the walls of the tanker and the sides of the bottle moved in, as if a vacuum were created on the inside.

When a model of air pressure is shared showing that air molecules can only push and not pull, this opens a meaningful discussion about sense-making. Why have we, as functioning adults and science educators, used suction to explain so many things—drinking out of a straw, using suction cups to attach things to walls, or using a vacuum to clean dirt off the floor? The answer is because that model seems to work well in explaining how we are sucking the beverage into our mouth, or how the vacuum seems to suck up the debris from the floor. But science doesn't suck. Air can only push or bounce off of other things. In this case, and in many others, we have not realized that this model of air does not work because we have not encountered a phenomenon that allows us to, as the definition describes, "ascertain the mechanism underlying a phenomenon" (Odden & Russ, 2019). Mechanistic explanations of phenomena require a step-by-step breakdown of what is causing the phenomenon to occur, in this case, breaking down what is causing the walls of the tanker and bottle to collapse over time.

The use of phenomena helps to uncover our current understanding of what we *think* is causing something to happen, while also illuminating the gaps in our own understanding. The use of phenomena helps us to say, "Well, wait, I don't entirely know why *that part* happens." Without encountering rich and complex phenomena, we can still easily move forward by explaining everything with a model like suction, that is, a model that works at a basic level, but that does not stand up to the science of what actually happens in the real world. As the British statistician George E. P. Box famously said, "All models are wrong, but some are useful." Suction as a model is wrong but useful, as millions of people use it daily to explain phenomena in their everyday lives. Raising these everyday explanations (even wrong ones) is important if we, as science educators, want to be able to formally assess where our students are in their current understanding, then guide them to engage in more complete explanations for what is happening in the world. Rather than seeing certain ideas as wrong and needing to be corrected, investigating these ideas can be seen as stepping-stones to deeper understanding (Campbell et al., 2016).

HELPFUL MISCONCEPTIONS?

The definition of sensemaking provided by Odden and Russ (2019) also highlights the fact that explanations are built "out of a mix of everyday knowledge and formal knowledge by iteratively proposing and connecting up different ideas on the subject" (p. 192). Everyday knowledge, or our intuitive knowledge, strongly shapes science learning and should be leveraged when working to develop deep conceptual understanding. From this approach, misconceptions are helpful to sensemaking in laying out a path to our own understanding. These ideas can come from our own experiences or that of others who are sharing their own thinking about why something is happening, or what they think may be involved. This *connecting up* of different ideas is at the crux of what true sensemaking looks like in our classrooms. Many students (and teachers) are frozen by their fear of saying the wrong thing while, in fact, scientists are full of conjectures about what "could" or "might" be causing something to happen.

For example, when a Slinky is fully extended and held out at arm's length and dropped, the bottom of the Slinky appears to remain in place as the top of the Slinky falls. Only when the top of the Slinky falls to meet the bottom does the bottom begin to fall to the ground. When showing this phenomenon in a middle school classroom, students have often shared that this makes sense to them because the bottom is not impacted by gravity just yet, or that the center of mass is why this happens. These are both really complex conjectures about a pretty challenging phenomenon, making it difficult to explain. With the initiation-response-evaluation model (IRE) of classroom discussions, a teacher might say, "Now, can anyone else think of what is happening here?" This shuts down these important lines of thinking and, in essence, wipes them off the board of possibilities for what could be occurring.

If this Slinky discussion were happening at the beginning of a lesson or unit, a more productive direction might be to make sure you are clear on what the student(s) are saying and capture that line of reasoning as one worth pursuing (i.e., writing it on the board or having students write it in their notebooks). Then a teacher's response purposed to elicit student sensemaking—such as "Yeah, I see what you are saying. Okay, what else could it be?"—brings other students into the discussion to try and share other productive ways of explaining this phenomenon. Only later will students have the opportunity to plan and carry out investigations (SEP 3) or obtain information (SEP 8) that might help them to meaningfully revise their thinking based on new evidence. This is fundamentally different from simply telling students that they are wrong and giving them the right answer. Research has shown that such facets of understanding constitute a more productive approach to supporting students in building knowledge over time (Lovett et al., 2023; Minstrell, 1992).

As a point of comparison, if you demonstrate or show a phenomenon to a scientist, you will almost always get one of two responses:

- They have seen it (or something like it) before and begin to immediately explain it.
- The phenomenon is new to them, and they begin searching around in their brain for resources they can use to make sense of it.

These two approaches are important to note because, as educators, we often find ourselves in these situations with students. Hopefully, the phenomena we are choosing to show to students are relevant in that they elicit intrigue and, yet, are complex enough to make students begin to explore what resources they have to explain them. While students are all different, one common way to tell if this is happening is when students have the "huh?" look, or when they are looking up to the left or right. I remember learning at one point that when you are looking up and to the right, it means that you are remembering something. On the other hand, when you look up and to the left, you are imagining something. While I am certainly not an expert on social cues, I know that good phenomena often get students to look up and dig deep into what they have seen and know about a topic. This is the starting point for sensemaking. Digging into that bucket of "What do I know about this?" means that students are starting to piece together an initial explanation. These initial explanations are usually messy, full of gaps, and benefit from being shared publicly and discussed. Labeling them as *wrong* shuts the door for sensemaking, especially as science in the real world is often messier and more nuanced than presented in many K–12 classrooms.

MESSY REAL-WORLD DATA

For many scientists, much of the day-to-day work is trying to make sense of messy, real-world data related to what they are studying. For example, data can look very different depending on the phenomenon being investigated—from studying insects in the rainforest, to studying the strength of different materials used to make plane engines, to analyzing massive amounts of global climate data. Nonetheless, the aim is the same—to analyze and interpret the data (SEP 4) in order to understand real-world phenomena. In all of these examples, there is no right or wrong way to approach making sense of the data collected. Based on prior knowledge, scientists might notice anomalies that cause them to ask questions (SEP 1) about one particular data point. Or they might see a trend that does not match current patterns. When students are working with data and evidence, they are often asked to make sense of it as a means to finding an answer. For this reason, students are often nervous in their efforts to interpret data on a graph or chart. A sensemaking approach shifts the aim beyond finding a single answer and, instead, has students looking into the data with the ultimate goal of understanding what they are looking at and what they can infer from the data collected. This is a fundamentally different thought process than peering at a table or graph included in a textbook to illustrate a relationship or demonstrate a trend.

The challenge to make sense of data for the general public has been shown over and over again in the real world. Whether looking at data related to climate, the pandemic, or even simple charts displayed on the news, findings tend to be often oversimplified and inaccurately summarized. We want our students to be scientifically literate and engaged sensemakers who can spend time analyzing data and coming to their own conclusions based on evidence. Why do the data make sense? What connections are the students making? Why does the sensemaking process seem biased or inaccurate at times? What additional data or information would help students make better sense of the data? Maybe the graph is showing that global average temperatures remained the same from one span across a few years. What would you expect to see if you zoomed out and looked at global average temperatures across the past 100 years? And what might you expect to see if you zoomed in and looked at hourly temperature data for one location across only a month's time? Whether developing a hunch or the ability to reason based on evidence, our aim is to work with real-world data and prepare students for success far beyond their next assessment.

One approach to bringing this kind of sensemaking into our classrooms is to work with real raw data. What I mean is data that have not been *cleaned,* a term data scientists often use to describe removing or clarifying their data to focus on only certain variables and/or time frames. Data that students are often working with are usually already filtered down to demonstrate a clear effect or relationship that can be observed. While this filtering can be important due to time constraints or in making sense of truly complicated phenomena, students do need to be given opportunities to make sense of raw and often messy data.

Over the years, I have worked with many students who tended to give up when given data to examine, saying that they "do not understand." For example, during my 3 years as Scientist in Residence at the Charles H. Barrows STEM Academy, a new K–8 magnet school, one of our goals was to immerse ourselves in the NGSS by pre-adopting the standards 2 years before the state of Connecticut ended up officially adopting them. In our work on analyzing and interpreting data (SEP 4), we found that students often gave up quickly because they "did not understand." Rather than giving students data and graphs that were easier to understand, we gave them data that were harder to understand, much messier, and more complex, but we *also* gave them context. This meant reaching out to scientists, via email, who were also studying the topics that we were studying—either at the nearby University of Connecticut or online—and asking these scientists to share with us the same raw data with which they were currently working.

You would be surprised how excited scientists are to share their data! Many even offered to meet virtually with students to share how they collected the data and to answer students' initial questions. The most exciting part of this process was seeing the shift in how students approached the new data. They knew the data were messy and harder to understand. At the same time, they saw the work as more authentic and different from *playing school* and often joked, "If we figure it out first, will the scientist put our names on the published paper?" I have shared this approach with many different schools and age groups and have seen similar results demonstrating that a shift to sensemaking, using real data, changed how students viewed their own work. They were more engaged and committed to making sense of the data on their own when they truly believed that there was no *right or wrong answer* hidden in the task. A shift from *learning about* to *figuring out* requires that there be something meaningful to figure out in the first place. But how do we organize this learning?

KEEPING LEARNING COHERENT FROM BOTH THE TEACHER'S AND THE STUDENTS' PERSPECTIVES

The very idea of sensemaking is enough to make any teacher or administrator nervous, as the process implies a degree of chaos in the classroom. How can a class behave if nothing makes sense and everyone is trying to figure out what is happening all the time? Luckily, a lot of research has gone into how to design learning experiences that are coherent from the student's perspective. For example, the wonderful team of Mark Windschitl et al. (2020) at the University of Washington captured a lot of this wisdom in their book *Ambitious Science Teaching* (AST; Windschitl et al., 2020). The AST approach is organized around four sets of core teaching practices: (a) planning for engagement with big ideas, (b) eliciting student thinking, (c) supporting changes in students' thinking, and (d) drawing together evidence-based explanations. In addition, the incredible team of Brian Reiser at OpenSciEd (https://openscied.org/why-openscied/design-specifications) have been working on open source curricular materials focused on a number of different grade levels and topics. These units are based on design specifications that follow an instructional model and classroom routines to meet the vision in the Framework and NGSS. I was excited to be a part of the team who wrote Chapter 1, "Instructional Model," and Chapter 11, "Classroom Routines," with Brian Reiser as the lead, along with Cynthia Passmore (UC Davis), Pam Pelletier (Boston Public Schools, retired), and Mark Windschitl (University of Washington).

HOW TO USE THIS BOOK

This book is organized to help teachers design, teach, and assess sensemaking-rich lessons. It's best if the reader reads the whole book all the way through before diving into planning so that they understand the whole process before they begin designing their own lessons. In Chapter 1, we'll talk about establishing the norms and values necessary for creating a culture that fosters sensemaking. In Chapter 2, we'll take a look at strategies for designing sensemaking experiences in elementary, secondary, and out-of-school settings. In Chapter 3, I discuss the teaching shifts and pedagogical strategies necessary for creating and leading sensemaking experiences. In Chapter 4, I walk you through the process of assessing sensemaking so as to ensure effective student learning and to inform future instruction. In Chapter 5, we'll consider ways of developing teacher capacity for sensemaking and the kinds of changes to preservice and inservice teacher education that will be necessary to equip all teachers to incorporate sensemaking in their science classrooms. Finally, in the Conclusion, I position sensemaking as the foundation for an overall shift in science education, aimed at ensuring that all students are engaged and supported. Essentially, I contend that the sensemaking approach to K–12 science education promotes lifelong learning and will ultimately result in a more scientifically literate citizenry—and it makes STEM a whole lot more fun for everyone involved.

CHAPTER 1

Getting Started

How Do We Create a Culture of Sensemaking?

The environments of sensemaking classrooms—in which students are mentally and physically engaged—are typically very different from traditional classroom settings in which students sit quietly in rows, focused on the teacher leading the lesson at the front of the room. My former principal at the K–8 STEM Academy would always jokingly say, "I can tell the room that TJ is in because it's the loudest class in the hall!" This comment always made me proud because the volume level the principal was referencing did not come from my voice; rather, it reflected the animated voices of my students going public with their ideas, arguing with each other by using the evidence from their science notebooks, and working together to figure out the best possible explanations for the phenomenon we were trying to figure out.

The dynamic kind of classroom culture associated with sensemaking does not just happen. It takes the establishment of clear norms specific to how we conduct ourselves in a space where we are working together to figure out science ideas and solve problems that we care about. While there is no shortage of examples and in teacher preparation programs concerning how to establish these norms online, it still remains one of the most challenging aspects of teaching and requires careful cultivation of classroom behaviors that support learning that includes sensemaking. In turn, it is also worth noting that when you hear scientists or engineers working together, you often notice these same elements of culture. To illustrate, in a meeting focused on installing a new tropical butterfly garden at the Connecticut Science Center, I remember sitting around a conference table discussing the environmental needs of the space. Engineers and other experts were arguing about the HVAC (heating, ventilation, and air conditioning) needs of this new tropical space. Each person had a different perspective because each brought a different or unique skill set to the group. For example, one person had created a model of the garden and was trying to figure out where to place the HVAC units so that they did not impede movement in the space. Another was thinking about how much energy was needed to power these units. Yet another had calculated the average temperature needs of a tropical butterfly garden and was working to calculate how much heating and cooling would be needed to reach those temperatures.

At one point, one of the HVAC experts looked at me, the entomologist on the team, and said, "How much are these butterflies going to be breathing in this space?" Because I was immediately impressed by this great question, I paused and asked the group participants to stop for a minute and think about the questions and issues each were aiming to address in this meeting. Were we trying to calculate how much butterfly respiration was occurring in an enclosed butterfly garden? How warm could we make an indoor greenhouse? No, we were trying to figure out the ideal temperature range for the humans (and butterflies) who would be visiting the garden. At the core, this was a phenomenon related to kinetic molecular theory—how much energy we would need to get these air molecules moving in a closed system. At the same time, such questions also involved the comfort levels for two different species (butterflies and humans) who have very different physiological needs in order to feel comfortable. The engineering challenge was to create an artificial ecosystem where both of these species could survive. At its core, this meeting was a sensemaking discussion in which the members of the group were trying to come to a consensus on a solution that would work based on our criteria and constraints (i.e., budget, space, access to power, needs of humans versus butterflies, etc.). The experts seated around the table participated in interactive discussions that reflected sensemaking practices informed by the NGSS (2013, Appendix F): (a) asking questions (practice 1); (b) arguing from evidence (practice 7); (c) using math and computational thinking (practice 5); (d) using models (practice 2); and (e) obtaining, evaluating, and communicating information (practice 8)—all without explicitly knowing anything about which NGSS practices they were using. In this meeting, each member had their own expertise, but reframing our collective goal to the group helped us to develop a shared language moving forward.

Most importantly, the meeting culture established by this group of engineers was one in which everyone's ideas needed to be heard and valued due to the diverse backgrounds and perspectives of each of the members. If even one person's thoughts were not heard or respected, the success of the group would be hindered. A classroom culture centered on sensemaking is very similar. Building knowledge in classrooms requires all students to have a voice in what they are observing, wondering, and thinking. Moreover, their multiple ideas and reasonings need to be made public so that all students can take in the value of wrestling with science ideas. This includes discussing how and why science and engineering practices will help them to engage and figure out more information, all while using the crosscutting concepts to help them think through different lenses in their sensemaking. It is no coincidence that sensemaking classroom spaces are often louder than traditional classrooms, where teaching is conducted in more passive ways. Thus, a sensemaking classroom involves students working in dynamic groups, students who are often at different places in their journeys of "figuring out" science ideas and solving problems.

HOW TO INTRODUCE SENSEMAKING IN CLASSROOMS

Research has shown that while teacher involvement is positively correlated with students' on-task behaviors, this support turns out to have a negative effect on student learning when the teacher takes on a role with too much control (Van Leeuwen & Janssen, 2019). So, how do we walk this fine line? How do we ensure that we do not take away our students' opportunities for learning while still making sure we provide the support they need (e.g., prompts, probing questions, just-in-time instruction, etc.)?

The approach I use starts on day one. I've found that the best way to introduce sensemaking is to immerse students in this kind of iterative thinking right away. Over the last 12 semesters, I have been teaching both Elementary Science Methods and Secondary Science Methods courses at Boston University, and we begin with sensemaking. In my Elementary Methods course, I start our sensemaking process right away with something called Euler's Disk. I do not introduce it by name, but instead say, "Okay, so I thought we should start with this cool thing I found." I show our preservice students a plate-sized (9″) mirror that looks like a makeup mirror (i.e., concave) and a metal disk that is very heavy for its size (3″ diameter, ½″ thick). After passing around the disk, I tell students that I want them to tell me everything they observe when I spin the disk on top of the mirror. Upon placing the disk on the mirror, the act of spinning it like a top causes the disk to rotate. Students typically watch intently at first, sometimes pointing out that it is spinning and rotating around the mirror before beginning to look up at each other to see if they are missing something. Initially, I stay quiet. The disk rotates around and around, spinning and making small concentric circles around the center of the mirror. It also makes an interesting sound and seems to be speeding up. Sometimes students will point out the sound. Other times, I will prompt students by saying, "Share anything you are observing—with your eyes, ears, anything." Then they usually say something like, "It sounds like it is speeding up." This observation is the beginning of something really productive! How does a sound convey a speed? And how can we hear the acceleration of a spinning metal disk?

Rather than jump right into prompting students to say more about what they mean, I've found that it is more important to position yourself as a part of the collective sensemaking process, saying something like, "Right, I totally hear that!" This kind of response, one that does not correct students or begin to explain what is happening, often creates a natural sense of wonder, and the ideas start flowing. For example, students begin to notice all sorts of things, including the movement of the spinning disk, the appearance of it seeming to leap up off of the mirror, and how it keeps spinning for much longer than they had initially expected. Once students have had a chance to share all of their observations, I move our group sensemaking into a second—and crucial—area. I ask students what this phenomenon involving the mirror and the spinning disk reminds them of. In other words, what kinds of personal experiences have these students had that they are able connect to this phenomenon? Students often share that this reminds them of helicopters, or more specifically, a video they saw where a helicopter is flying but the blades do not seem to move in the video. Or it makes them think of the rims spinning on cars on the highway. Whatever students share, these related phenomena help to provide a window into students' sensemaking, even providing a sort of formative assessment that allows you to see how they are connecting this new phenomenon with other things they have experienced in the past.

I had learned of the importance of *related phenomena* from Brian Reiser, a professor of Learning Sciences in the School of Education and Social Policy at Northwestern University, who was then doing research on high-quality NGSS curriculum. His team noticed that some teachers using early versions of the OpenSciEd curriculum were having more success than others. To figure out why, the team visited classrooms and noticed that the teachers who were having more success were not just moving directly from observations to questions (when experiencing phenomena); at some point, they asked students what the phenomenon reminded them of from their personal perspectives. In other words, the teachers did not voice this prompt immediately. They first encouraged students to share their initial observations and questions. It was only after students shared most of their initial ideas that the teachers prompted them to consider connections between their own experiences and the phenomenon under study. By asking students to think about and make connections in this way, the questions stepped up a level and often became more aligned with the disciplinary core ideas (DCIs)—not just about the *thing* they were observing.

For example, consider the middle school instructional unit on sound. The phenomenon is basically a record player made out of a spinning platform, a sewing needle, and a paper cone. Students, many of whom are unfamiliar with traditional vinyl records, are surprised to hear that moving the needle over the disk (the record) causes sound to become audible. Instead, many students' initial questions tend to center on the physical phenomenon they are observing. What is the record made of? What is the needle doing? What does the paper cone have to do with it? While these are important questions, there are no DCIs related to how record players work. However, when the teachers ask students to think about what the record player setup reminds them of, the observations and questions tend to shift to deeper questions that connect to the DCIs. This scenario reminds me of hearing a plane before I see it, hearing music when I am swimming underwater, and driving over the musical roads (i.e., rumble strips that make sounds in the car as we drive over them). These sensations represent additional examples of how the interaction of a physical thing causes sound, how sounds can move through different media, and how they eventually reach our ears.

Broadening the scope of study to other related phenomena is crucial to reaching the core ideas in the NGSS for many reasons. For one, doing so reduces the pressure of depending on a single phenomenon to carry the weight of an entire instructional unit. A broadened scope further validates students' ideas and allows them to authentically use their own prior experiences in the STEM classroom. This approach also brings students' questions and partial ideas forward in a way that allows for formative assessments. And, probably most importantly, it makes the classroom a place where student sensemaking *requires* students to share their thinking out loud in a learning community in which sensemaking is a collective act.

THE COMPLEXITIES OF PHENOMENA

Since the launch of NGSS Phenomena (https://www.ngssphenomena.com) in 2015, I have been involved in many conversations about what is and is not a phenomenon. Because I have seen numerous lists of presumed phenomena and have had so many educators reach out to me to verify if something *counts* as a phenomenon, my thoughts on this topic have brought me to a very simple conclusion: If the subject/thing is academically productive for you and your students to investigate, then I believe that it should count as a phenomenon worthy of study. If it inspires wonder in your students, they will start asking questions and want to figure out the science ideas that are necessary to explain the phenomenon. At this point, you are in the right place! Trust yourself and your own expertise about what is best for your students. I find that in the majority of cases, the idea just needs to be refined a bit to figure out what helps you to best engage your students and help them reach your learning goals.

Yet many science educators still wonder about what qualities or characteristics constitute a phenomenon, so I want to focus on what makes something more (or less) academically productive as a phenomenon for science teaching and learning. First, the phenomenon does not need to be *phenomenal*. This means that it does not have to be something over the top like an explosion, or a volcano erupting, or a video of a huge science demonstration, for it to be effective. A productive phenomenon is anything that allows students to engage and inquire about the scientific ideas behind its occurrence. Note that there is an important distinction to be made between something that engages students in figuring out science ideas and something that students are interested in that *does not* help them understand core science ideas. For example, students might be very interested in a unit requiring them to imagine an animal that lives on another planet, including creatively describing the animal's body features and how it survives. While this activity can serve as an engaging exercise, if it does not lead students to an understanding of actual organisms that exist on planet Earth, then this approach will not be academically productive.

A second important element of science education is understanding that the study of a phenomenon is not considered productive if it requires that the science ideas be pretaught in order for students to make sense of them. There is nuance here, because as students figure out more and more core ideas in science, they will certainly be able to figure out more complex phenomena. In addition, "Just-in-Time Instruction," from Ambitious Science Teaching (AST; https://ambitiousscienceteaching.org), is a way of strategically introducing conceptual ideas at moments when students have a need to know more, and helps to bridge the gap between their current understanding and the information they need to move forward in their learning. The Just-in-Time approach, to be further discussed in Chapter 3, supports students in figuring out more complex phenomena, but its process is not the same as actually preteaching science ideas before introducing a phenomenon. Essentially, investigating a phenomenon should cause students to want to understand the science ideas underlying why and how it occurs, thus allowing for a carefully constructed unit of instruction intended to support students in figuring out important science ideas over time. While complex, a sequence of teaching events that introduces the phenomenon first enables students to develop their own understanding of science ideas rooted in exploring the phenomenon itself—in contrast to learning facts and other decontextualized science ideas and then trying to apply them to a particular scenario.

A third important area of critique regarding K–12 science education is the idea that all phenomena need to be things that happen in the classroom. We know that not every science idea can be easily replicated in the classroom. According to Ambitious Science Teaching, events happening outside the classroom can still be academically productive as long as they are context-rich and focus on a specific event at a specific time. For example, the AST site includes an example led by Bethany ("High School Series: Ecosystem Dynamics") in which students were trying to figure out how to answer the following question: *Why, in a local ecosystem, was the population of hares oscillating up and down every seven years*? The students started their investigation by sharing what they thought would happen to the population over the years. While this phenomenon does not represent something happening *in* a classroom, it is a great example of something that does happen in the world and that students can figure out through the use of the science practices and crosscutting concepts.

Fourth, a number of the criteria lists state that a phenomenon is an observable event that happens in the real world. This is an ideal definition, since a phenomenon is easier to make sense of when all students can observe it in real time, together, as a classroom community. The observable study of a phenomenon is most helpful in creating engagement, crucial for promoting access and equity by focusing on things that are relevant to all students. Yet the world of unobservable events constantly happening around us can also lead to productive sensemaking. Therefore, illuminating such unobservable events through engaging and accessible phenomena should remain one of our goals in the field of science. For example, I had an interesting conversation at the Bigelow Laboratory of Ocean Sciences with senior research scientist Nick Record. We were looking at a recent line graph of climate shifts, and Nick pointed to a spot in the graph where the line was moving up at an aggressive slope and then, all of a sudden, shifted down. He shared that as a research scientist, movements like this in the data tell him that something needs to be explained, that a mechanistic explanation must exist for why the numbers were going up, stopped, and then started going down. While this scientific approach should seem straightforward—we are always trying to make sense of data—I took Nick's comment as a great example of one of the crosscutting concepts: cause and effect. Something was causing the numbers to rise (line graph going up), but something else caused that to change and the numbers to go down. The effect that Nick was keying in on was the downturn in the graph, which does represent something important happening in the world. Still, what specifically caused that change in direction has not been figured out.

A fifth area of complexity in science education relates to the idea that a phenomenon needs to be a single thing that can be figured out. In reality, a single phenomenon will rarely allow you to unpack enough science ideas for it to be academically productive. Rather, a series of phenomena need to come together in a way that allows students to get at the broader core ideas in science. A great example of this is the new NGSS badged instructional unit provided by Vivayic, a learning design company, titled "Media Mayhem." This unit explores claims made by media outlets across the country—both positive and negative claims—about the impacts of milk production. In this case, the phenomenon is not specifically how milk is produced. Rather, the phenomenon refers to the varied and competing claims about milk production being either a good or a bad thing; as such, leading students to uncover the components of how the dairy food system impacts the environment (i.e., pollution, greenhouse gas emissions, climate, biodiversity, etc.). These impacts signify core ideas that students need to figure out in high school, using the ideas to engineer multiple solutions that might reduce the negative impacts of the dairy industry. This unit is an example of how a phenomenon can consist of multiple and different aspects, yet, when brought together in a careful manner, these multiple aspects can provide students with an avenue to make sense of many core ideas using the NGSS science and engineering practices and crosscutting concepts.

The last distinction that I feel is important to make is the difference between unit-level or anchoring phenomena and lesson-level phenomena. A unit-level phenomenon is something that students can keep returning to after each lesson, adding more to their understanding as they incrementally learn more and more about the science ideas. A lesson-level phenomenon is smaller and easier to figure out in one or a few lessons that contribute to an overall understanding of something larger. For example, in New York University's Science and Integrated Language (SAIL) unit "Garbage Unit" (NGSS badged unit), the anchoring phenomenon is the huge amount of garbage created each day by schools, homes, and neighborhoods. Consideration of this anchoring phenomenon leads to the driving question, "What happens to our garbage?" This is the question that students return to after most lessons so as to see what they can add to their prior explanations. Early on in the unit, students tend to think about how landfills work and if/how the discarded materials are changing in the landfills. Once students determine that those materials might be changing, they bring up some of their prior experiences with smelly garbage and decide to investigate smells. In Lesson 3, students investigate a lesson-level phenomenon focused on the question, "Why can I smell the pizza in the lunchroom from the hallway?" This is a lesson-level phenomenon because it helps students to make sense of smells and air (matter) and then calls for them to apply this new knowledge back to what is going on in the landfills.

A helpful note here is that when you call for students to share what a phenomenon reminds them of, they will often share a lot of really interesting examples that connect to other phenomena you have in your unit. Write these down! Recorded notes can be very useful, providing you with an incredible list of candidate phenomena to be used when revising your units in the future. I created a Google form for my own feedback that only asks four questions: (a) What unit was I teaching? (b) Which lesson? (c) What went well? and (d) What could be improved in the future? Filling out this form only takes me a few minutes max, and the process allows me to do something that I always tell students to do—be nice to your future self. As a science educator, I have learned that recording things in the moment is invaluable when you sit down later in the year or over the summer to make improvements to your curriculum for upcoming classes. Also, do not be too hard on yourself. Whether you are an elementary teacher who has little time after teaching science to reflect before moving on to another subject or you are a secondary teacher who has 100+ students and back-to-back classes, giving yourself feedback even once a day, once a week, or a few times a month can have a huge impact on kickstarting your own curriculum work later on. In addition, taking the time to review your notes when sitting down with a STEM coach or curriculum team is a great way to keep your unit revisions focused on evidence and actual experiences with your own students—and not trying to remember how things went way back when you taught it previously.

Overall, the topic of phenomena for the NGSS is complex because it is often one of the first takeaways for science teachers when being introduced to the new standards. Nonetheless, as you learn more and begin implementing these practices in your classroom, a whole new world of possibilities opens up. In terms of incorporating the new standards, the first urge for the science educator is usually the desire to search for new phenomena for everything they are already teaching. This takes a huge amount of time and can leave you feeling overwhelmed. Rather than trying to overhaul everything, start with one unit and begin to make incremental changes. Adding a phenomenon should be an exciting place to start and not a stressful moment where you feel like you need to get it right the first time. I often think about the quote from Stanford University's Dr. Bryan Brown: "We are not teaching science, we are teaching students, and we want them to learn science." To me, this means that the phenomena alone will never be perfect from year to year; rather, they give us the creative liberties to adjust our teaching to truly meet the needs of all of our students.

THE CHALLENGES OF SETTING NORMS FOR PRODUCTIVE SENSEMAKING

For meaningful learning to take place, it is crucial that the science educator create a classroom environment that supports risk-taking by encouraging students to share their ideas openly and that fosters the understanding that no idea is ever final. This kind of atmosphere doesn't happen by accident; it requires the intentional establishment of norms and values that champion these principles. When students walk into a space where they feel safe to express their thoughts, no matter how incomplete or experimental, they begin to see learning as an active, iterative process. They understand that making mistakes and revising their thinking are aspects of learning not just allowed, but that are also essential parts of the journey. In my experience, the most vibrant and effective classrooms are those in which students feel empowered to test their hypotheses, argue from evidence, and collaboratively refine their ideas. This dynamic environment is underpinned by the belief that learning is not about getting the right answer on the first try; rather, learning is about exploring possibilities and refining understanding over time. By embracing the idea that no concept is ever truly complete, we can cultivate a classroom culture in which students are not afraid to be wrong. Instead, they see errors as valuable stepping-stones toward deeper insights. That said, creating this culture is not easy.

For a sensemaking culture to thrive in the classroom, it is vital to set clear norms from the outset. These norms should encourage respectful dialogue, critical questioning, and the view that each student's contribution is valuable. When students feel their voices are heard and their ideas respected, they are more likely to take intellectual risks and engage fully in the learning process. In my own classroom, whether we are debating the environmental needs of a butterfly garden or unraveling the physics behind Euler's Disk, the focus is always on collaborative sensemaking. We build our understanding together, continually testing and refining our ideas. This approach not only mirrors the practices of scientists and engineers working in the field, it also prepares students to navigate a world in which knowledge is constantly evolving. By fostering an environment where risk-taking is normalized and iterative learning is celebrated, we can equip our students with the mindsets and skills they need to tackle complex problems both in and out of the classroom. So let's create educational spaces where curiosity can thrive, where students feel confident to speak up, and where the journey of learning is as valued as the destination.

As an example, a useful approach that has been productive in creating this type of culture is one we use in our NGSX professional learning sessions. In working with adult educators, we model what a productive norm-setting conversation might look like in the K–12 classroom. We start with a poster that already has the following items or characteristics of productive conversations listed on it:

- Talk that is . . .
    - » Equitable
    - » Respectful
    - » Focused on reasoning

We begin with these three characteristics because they represent important starting points. At the same time, we leave space under each bullet for the group to decide how they would like to collectively define what each term means to them. Turning to the group, we ask, "What does it mean to have equitable discussions?" Someone in the group might offer, "It means everyone gets to speak." However, before writing that down, we might ask, "Gets to, or has to, or can?" How should we consider the role of our students sharing in our discussions about equity in classroom discussions? Is an equitable discussion one in which everyone has to speak? How do we make sure that everyone's voice is a part of every discussion? How do we balance the dynamic of the student(s) who always has something to say right away with the student(s) who needs some time to develop their ideas and/or questions? Taking time to have these conversations results in a productive dialogue about why everyone *needs* to be heard during sensemaking. We follow the same steps for addressing each point on the poster, making sure that we are pushing the conversation beyond a few simple words and toward a full description of what *talk that is equitable, respectful, and focused on reasoning* actually looks like in classrooms. When groups are having a tough time articulating what these norms look like in practice, it can be helpful to ask the group (teachers or students), "What does it look like when these productive characteristics are NOT happening?" In other words, non-examples are often helpful in illustrating what kinds of things we do want to see and what we are hoping to avoid in these discussions.

The process of creating this list together with your learning community is crucial to establishing norms upon which the whole class can agree.

Note that this list is not random. It does not include anything that educators and students share without being shared back with the group for their thoughts. Also, it is based on what we, as educators, know about talk that is equitable, respectful, and focused on reasoning. Facilitating these discussions is a careful balance between helping the group have ownership over their norms while also making sure to construct a list that will be productive for sensemaking. To support this endeavor, I next briefly address each of these points as characteristics of productive classroom conversations.

Equitable talk ensures that all students—regardless of their backgrounds, abilities, or confidence levels—have equal opportunities to participate in classroom discussions. This inclusivity allows for a rich diversity of perspectives, enhancing the learning experience and helping students appreciate different viewpoints. When every student has a voice, they build confidence in their ideas and contributions. This inclusive approach ensures that all students contribute to the collective understanding of the class. When students feel included and valued, they are more likely to engage actively and think critically, which enhances their ability to make sense of complex concepts.

Respectful talk is fundamental in creating a safe learning environment in which students feel comfortable expressing their ideas without fear of ridicule or judgment. It promotes constructive dialogue and encourages students to listen to and consider the viewpoints of others, leading to meaningful and productive exchanges. Respect in discussions supports the emotional well-being of students, fostering a culture of mutual respect and empathy. When students respect one another's contributions, they are more likely to engage in deep, thoughtful analysis and build on one another's ideas—all leading to a richer understanding of the material. Respectful talk ensures that discussions are focused on understanding as opposed to personal attacks, thus making discussions more engaging and productive.

Finally, talk that is focused on reasoning emphasizes the importance of evidence and logical thinking in discussions, helping students develop critical thinking skills. This approach encourages students to articulate their thinking clearly and support their ideas with reasons and evidence, promoting clarity of thought. Additionally, reasoning helps students develop problem-solving skills by teaching them to approach problems systematically and consider multiple solutions. When talk is centered on reasoning, students learn to value the process of thinking through problems and explaining their reasoning. This not only deepens their understanding of content, but it also fosters a culture of inquiry and intellectual rigor—essential for effective sensemaking.

The integration of these three elements of productive norms—equitable, respectful, and a focus on reasoning—creates a dynamic and inclusive classroom environment that encourages all students to think deeply and share their insights. This kind of interaction enhances individual understanding and promotes building a collaborative learning community. When students learn from and with one another, they collectively develop a stronger ability to make sense of the material, fostering a culture of sensemaking that benefits everyone.

THE IMPORTANCE OF EXPLANATIONS IN SCIENCE EDUCATION

The Framework emphasizes the importance of explanations in science education, highlighting the key role of explanations as the means with which to help students develop a deeper understanding of scientific concepts. When students construct explanations, they make sense of phenomena and connect their knowledge in a coherent way. This knowledge increases in complexity as our students progress through their years of education, leading to more and more thorough explanations. In K–12 classrooms focused on sensemaking, explanations do not end at an *answer*. Rather, they build toward increasingly more complete explanations. This means that shifts are needed in many classrooms—away from explanations serving as shortcuts to the final answer or end goal. Instead, explanations need to be iterative in nature, involving the active construction and deconstruction of ideas as they are tested and refined over time.

How do we reach the goal of using explanations as drivers of ongoing inquiry in real classrooms, where both students and teachers have been conditioned to think of explanations and answers as the same thing? I'm sure many of you reading this right now can remember, as students, countless times when you answered a question on a quiz or test, followed by instructions to *explain your answer*. These two test responses—the answer and the explanation for it—have become intertwined in a way where explanations are thought to be the end and not a productive part of the sensemaking process. The Framework specifically highlights the following point: "To develop a thorough understanding of scientific explanations of the world, students need sustained opportunities to work with and develop the underlying ideas and to appreciate those ideas' interconnections over a period of years rather than weeks or months" (National Academies of Sciences, Engineering, and Medicine [NASEM], 2012, p. 26). As humans, we are always building explanations of phenomena in our day-to-day lives, and these naturally become more complex and interconnected over time.

To help us construct classrooms focused on sensemaking, I believe that we need to better understand our natural instinct to construct explanations. Professor Tania Lombrozo, director of the Concepts and Cognition Lab at Princeton University, has conducted extensive empirical research on the nature and function of explanations in the field of psychology. Lombrozo (2006) has argued that explanations help us make sense of new information, integrate that information with existing knowledge, and guide our learning processes. She has found that explanations often involve identifying causal relationships (Lombrozo & Carey, 2006); further, that "explanations are considered more favorable or beautiful if they are simple and broad in scope" (Lombrozo, 2018). This view of explanations aligns with the vision in the Framework for disciplinary core ideas (DCIs) in that both are broad and address multiple phenomena. In addition, Lombrozo and Wilkenfeld (2019) found that explanations serve pragmatic functions such as persuading others, justifying actions, or satisfying curiosity—depending on one's goals. Here, the science and engineering practices (SEPs) come to mind as the *doing* of science. We use the SEPs for pragmatic functions like (a) persuading others (e.g., arguing from evidence); (b) justifying actions (e.g., how we design and carry out investigations); and (c) satisfying curiosity (e.g., asking questions and constructing explanations). Why do we *do* science this way? Because we truly want to figure things out. Using curricula that engages students in authentic sensemaking encourages them to *want* to figure things out because doing so is satisfying. This notion is crucial to the work of teaching and learning because satisfaction has been found to predict children's preferences for and recall of explanations (Frazier et al., 2016). The sense of satisfaction impacts their motivation and perceptions of learning (Liquin & Lombrozo, 2022). Similarly, when I am motivated to figure something out, I perceive the learning process as rewarding and enjoyable. I simply cannot wait to figure it out, and I enjoy the process of doing so. In such moments, I find that questions flow naturally, and I go out of my way to ponder how I can figure things out. In other words, when someone gives me the answer to a particular question or Googles it and then tells me the answer, they rob me of that *aha* moment for which I am so desperately searching. Ultimately, classrooms grounded in a culture of sensemaking foster curiosity and support students in using the SEPs and crosscutting concepts to figure out phenomena. In this view, the challenge of solving problems sits at the center of everything they do.

While Lombrozo's writing can be complex and at times difficult to understand, I believe there are some key takeaways from her work that STEM educators can learn and apply to their teaching practices in K–12 classrooms. While the following is not an exhaustive list of takeaways, it can help teachers support students in their efforts to create the kinds of explanations that lead to true sensemaking.

1. **Encourage Explanation:** Prompt students to explain their thinking. This process helps students deepen their understanding and promotes critical thinking skills. Regularly asking students to articulate their reasoning moves them beyond rote memorization to a more nuanced understanding of concepts.
2. **Promote Self-Explanation:** Foster the habit of self-explanation, whereby students explain material to themselves as they learn. This technique improves problem-solving abilities and the transfer of knowledge to new situations. Incorporate activities that require students to reflect on and explain their learning processes.
3. **Focus on Reasoning Over Answers:** Emphasize the reasoning process rather than just arriving at the final answers. Design assessments and activities that evaluate students' thought processes and their ability to make logical connections—not solely their ability to arrive at correct answers.
4. **Value Collaborative Discourse:** Encourage collaborative discourse during which students discuss and explain their reasoning to peers. This approach enhances understanding and builds communication and social skills. Create a classroom environment that encourages respectful and equitable participation from all students.
5. **Use Diverse Examples:** Provide a variety of examples and scenarios to help students apply their knowledge in different contexts. This practice reinforces their understanding and ability to generalize concepts. Incorporate diverse and relatable examples to cater to different learning styles and backgrounds.
6. **Provide Feedback and Iteration:** Give regular, constructive feedback to help students refine their understanding and improve their reasoning skills. Implement iterative learning processes whereby students receive feedback, reflect on it, and make improvements.
7. **Harness the Power of Wrong Answers:** Encourage a culture that values "wrong answers" as opportunities for sensemaking. This approach centers evidence-based science and helps students explore diverse perspectives while also drawing on their own experiences.

A classroom culture conducive to sensemaking is not easy to establish. Nonetheless, a sensemaking environment provides the foundation upon which learning will flourish. Students will be more motivated, engaged, and cooperative in working together, and they will enjoy the process of figuring out phenomena and solving problems. Now, how do we design units and lessons where sensemaking can occur? In the next chapter, I focus on strategies for designing learning experiences that foreground sensemaking in elementary, secondary, and out-of-school settings. While the content of Chapter 2 is not meant to serve as a comprehensive education on curriculum design, I provide a series of strategies and practical examples that will prepare you to start integrating sensemaking into your classroom right away.

CHAPTER 2

Designing Learning Experiences That Foreground Sensemaking

Designing three-dimensional sensemaking lessons requires an intentional and thoughtful approach that leverages science and engineering practices (SEPs) and crosscutting concepts (CCCs) to uncover and explore core disciplinary science ideas. Unlike more traditional science lessons, which often focus on confirming lab findings or reinforcing pretaught concepts, sensemaking lessons engage students in the active construction of knowledge. This means creating opportunities for students to authentically grapple with real-world phenomena, ask questions, and use evidence to build their understanding. The key question is: How do lessons grounded in sensemaking differ from more traditional instructional methods that are more passive in design, that is, not actively involving students in figuring out key ideas?

One key aspect of designing sensemaking lessons involves the integration of science practices, such as asking questions, developing and using models, planning and carrying out investigations, analyzing and interpreting data, and engaging in argument from evidence. By embedding these practices into the lesson structure, students are not just passive recipients of information. Instead, they become active participants in the learning process. This kind of dynamic engagement fosters deeper understanding and retention of scientific concepts (DCIs) because students see the relevance of what they are learning in the context of real-world applications. In addition to science and engineering practices, CCCs like patterns, cause and effect, systems, and system models play a crucial role in sensemaking lessons. These concepts help students make connections across different areas of science (DCIs) and apply their knowledge to new situations. For instance, understanding the concept of cause and effect can help students make sense of why a particular chemical reaction occurs, while recognizing patterns that can aid in predicting future events based on observed data.

The shift from traditional to sensemaking lessons also requires a change in classroom dynamics. Teachers must create an environment in which students feel safe to share their ideas, make mistakes, and learn from them. This involves establishing clear norms and fostering a culture of respect and collaboration. Practical advice for teachers includes (a) starting with a phenomenon that sparks curiosity, (b) allowing time for students to discuss and reflect on their observations, and (c) guiding students to use evidence to support their claims. In turn, specific examples of sensemaking lessons can vary widely—from elementary school students investigating why leaves change color in the fall to high school students exploring the impacts of climate change on local ecosystems. In each case, the focus is on students actively constructing their understanding through engagement with a phenomenon, considering relevant problems and figuring out important questions and possible answers collectively, as a classroom community. By designing lessons that integrate science and engineering practices and crosscutting concepts, educators can create dynamic and inclusive learning experiences that prepare students to think critically and solve complex problems.

In this chapter, I begin with an overview of the important elements specific to the kinds of instructional units and lessons that foreground sensemaking. This approach was inspired by the many conversations I have had with incredible thought leaders in the educational space, conversations and sharing of information that I believe are crucial to establishing a foundational understanding of what is needed for high-quality instruction in STEM. Following the overview, I break down the practical advice needed to get started in building new sensemaking lessons or customizing more traditional lessons into ones that will provide each student the agency to take a lead role in their own learning.

INSTRUCTIONAL UNITS GROUNDED IN SENSEMAKING AND COHERENCE FROM THE STUDENT PERSPECTIVE

In a coherent storyline, the flow of lessons is designed to build toward meeting established learning goals in a way that makes sense to students from their own perspectives. Instructional units build toward the achievement of unit goals through the use of a storyline (i.e., a coherent sequence of experiences across lessons) that resonates with students, whereby each lesson is an extension of prior learning as motivated by questions generated to explain a phenomenon or to solve a problem. Most, if not all, questions stem from the original anchoring experience, the phenomenon under study, or ongoing challenges—often presenting new, puzzling elements that arise as students make progress in their explanations or design solutions.

In terms of planning and designing sensemaking lessons, thinking about coherence from the student's perspective is challenging for many teachers at first. Most preservice teacher education programs have taught us that as educators, we are the ones who know how to best guide student learning so as to meet each unit's learning goals. While the statement still rings true today, the NGSS calls for students to be the ones driving their sensemaking, and they simply cannot do this without educators who design and facilitate learning experiences in a different way. This new and different approach can feel uncomfortable for many educators, especially given the potential of the NGSS to "re-novice" the K–12 science teaching force (Hanuscin et al., 2018), that is, putting teachers with years of experience in the position of feeling underprepared all over again.

When I think of coherence from the student's perspective, I imagine walking into a classroom and asking a random student, "What are you doing?" If they respond, "We're learning about animal adaptations," or some version of a prescribed learning standard—which may or may not be posted on the board—then they are not doing the kind of learning that is grounded in sensemaking. Instead, if they respond, "We're trying to figure out why some animals have very different kinds of eyes and ears even though they live in the same kinds of environments," that level of response shows that they are fully aware of what they are trying to figure out. In this example, while both lessons could share the same learning goals, they are very different in how they are framed. Based on the second response, the student knows what they are trying to figure out, and we can infer their overall comprehension of where they are heading next in this unit. This kind of learning does not just happen; rather, it is carefully planned and orchestrated by the teacher. To achieve this kind of sensemaking response, the teacher needs to fully understand the storyline of the unit, including knowing how much content to introduce and how to present it so as to draw in their students' interest. This kind of teaching is not about the fidelity of implementation specific to each unit. Rather, the sensemaking approach gives teachers the agency to customize the learning pathway based on their own expertise regarding what their students know and what interests them. Additionally, this expertise grows as students begin to understand the phenomenon or problem. In other words, the science teacher's expanding expertise is informed by the ways in which students demonstrate how they are integrating the pieces of the puzzle—by comparing competing possibilities and assembling elements of a model, explanation, or design solution developed across the unit to evaluate its completeness. All of these are formative assessments that potentially lead to additional cycles of questions and investigation to form a more robust series of lessons.

At the same time, coherence can be viewed through an overly simplistic lens. An example of this would be a literal interpretation of the phrase *students drive all the learning*. This interpretation could leave teachers feeling pressed to go to Target every night to get new materials needed to address students as the drivers of learning. This is not what we are striving for. Teachers are professionals and know where students need to go in achieving their learning goals. So, how do we balance out this interpretation? First, we must acknowledge that the standards exist. Thus, we know what we need our students to know and be able to do by the end of our time with them. Still, there are ways that we can shift the balance so that students are involved in the curricular decisions in terms of moving in one direction or another. A lot of this approach has to do with *why* we should focus on the next topic rather than just addressing what we are going to do next and how. This means that instead of saying our next lesson is on photosynthesis, a teacher might say, "We know that trees get bigger, but if they are not absorbing the soil or being made up of fertilizer and water, then how could they be gaining all that mass?" The next natural step is to learn about the process by which the trees synthesize nutrients from the sun and air. However, the sensemaking approach frames the *why* in a way in which students become co-drivers in knowing the reason for learning about this process known as photosynthesis. The teaching moves required to emphasize the *whys* of learning are subtle, but they are powerful for encouraging and helping students leverage their own curiosity while also supporting them in using the science and engineering practices and crosscutting concepts as tools for working at knowing.

SENSEMAKING AS A SOCIAL ENDEAVOR

In many traditional classrooms, students often feel as if there are a limited number of points that can be handed out for each assignment. Therefore, they must individually compete to get more points than their classmates. This is an antiquated concept, and many updated grading systems are emerging (i.e., upgrading, standards-based grading, contract grading) in an attempt to recenter learning as the goal, as opposed to students simply going through the traditional, rote steps to get the A and usually forgetting a lot of the material they learned right after the test. These new grading systems do not mark the grade as the end goal of teaching and learning. Rather, the new grading systems have been implemented to increase student engagement, provide opportunities for meaningful teacher feedback, and create assessments that better align with the overall learning objectives.

One of the major shifts in a classroom that is centered on sensemaking is the focus on collaboration (i.e., collective sensemaking). With a collaborative mindset, students are not competing for points. Instead, students work together to truly listen to the ideas of their classmates and use those shared ideas to enhance their own reasoning. In sensemaking classrooms, diverse thinking is required because it expands each student's view of a phenomenon or problem. Further, the thinking of their peers helps to push their own thinking into new and exciting places. In order to successfully facilitate collaboration among students in the classroom, lessons should include small groups of students engaged in argumentation, explanation, and modeling. As in the whole-class group context, all members of small student groups need to be enabled to draw on their own life experiences in order to inform their own sensemaking capacities. For example, if 1st-grade students are trying to figure out *why some animals have very different kinds of eyes and ears even though they live in the same kinds of environments,* they might be introduced to a limited number of animals with different characteristics. At the same time, I would draw heavily on other animals that students are more likely to be knowledgeable about based on their own life experiences. More familiar animals might include family pets, animals from local zoos or aquariums, animals that students have read about in books and magazines, and animals that students have seen on YouTube or via other online resources. This approach not only broadens the perspectives of the classroom community, but it also brings individual relevance to their own schooling and helps students to see themselves as important members of the sensemaking community.

It is worth noting that not everything students do in a sensemaking lesson is done in smaller groups, yet with the whole class getting the same grade. There is nuance to be found in a lot of these instructional shifts. Students clearly need time to think individually and come up with their own ideas, questions, and observations. In the real world, ideas still need to be shared for actual progress in STEM education to occur. Whether you are collaborating with a team of engineers to design a butterfly enclosure, trying to determine if your scientific findings are new and applicable to the field, or working with a group that is developing a new technology, all of these endeavors require making your thinking public and using the science and engineering practices to reach a collective goal. This is the kind of environment that we, as science and STEM educators, are working to establish in order to prepare all students for engaged learning and successful educational experiences. As a matter of fact, preparing all students for success includes those students who do not have any desire to go into a STEM field. This perspective has become increasingly important as the speed of innovation and access to information indicate that students are constantly exposed to and trying to make sense of headlines from around the world, in particular via social media videos and news stories that are constantly changing. Whether it is a claim about some new solution to climate change, a graph showing a hard-to-interpret scientific finding, or a news article about an engineering event like the Baltimore bridge collapse, we need all students to be prepared to use developed STEM sensemaking skills in order to parse signal from noise. They need to be able to individually make sense of phenomena, share their own perspectives, and then take part in actively critiquing, defending, and negotiating their own ideas with others.

From a practical perspective, there are many tools you can use to incorporate collective sensemaking in your science classroom. The first tool, which might sound antithetical, is the *science notebook*. Used as a creative space for students to capture their ideas through writing, diagrams, and tables, science notebooks provide the space for students to think first—before bringing their ideas to a broader audience. Writing and diagramming in a science notebook is helpful for students who may not wish to be the first to share in a group setting. Yet, with ideas in hand (and in their notebooks), these students can reference their evidence, questions, or other ideas when opting to participate in group discussions. In this way, the use of science notebooks helps students to be more prepared to support a dynamic kind of group work that might be difficult without notebooking in place from the start. Science notebooks also provide the space for students to capture their thinking over an extended period of time. Ideas change and develop constantly when engaged in sensemaking, requiring a place to capture new thoughts or new questions that may be worth pursuing in the future. I always share with my students the science notebooks I used when doing animal behavior research to highlight possibilities and changes that I could add to my original notes. In other words, while there are a lot of ideas and questions in these notebooks, most are only partial ideas. The important aspect or benefit of using science notebooks is the dedicated space they provide us to record and see how our ideas (and questions) grow in complexity over time.

Another tool to support collective sensemaking is a *driving question board* (DQB). A DQB is a visual and interactive tool used to organize and track questions over the course of a unit. It provides a central location for student questions about phenomena or problems to be displayed in a way that drives the learning around answering these questions in each lesson. All DQBs generally start with a central question that the unit is focused on, and this leads students to contribute their own questions as they are generated during initial observations, discussions, investigations, and additional activities that spark students' curiosity. As a classroom community, these questions are then organized into themes or subtopics that help to guide the learning process. A key element of the DQB is that it is an active part of the process, not just a classroom artifact that is created once. Rather, the DQB is updated continuously throughout the unit as students answer some questions and also come up with new ideas and questions during the unit. Overall, the DQB supports group sensemaking because it is a public display of all of the questions that students and their peers have created in relation to a phenomenon or problem under study. Essentially, the DQB is used by groups of students to organize and drive their learning throughout a unit of study.

An equally important tool in supporting collective sensemaking is the *summary table*. While there are many versions of this tool, the general structure is a table with columns labeled according to (a) what we did, (b) what we observed, (c) what we figured out, and (d) how our ideas relate to the original phenomenon or problem we are addressing. As positioned, the first column enables you to capture the topic of the lesson (*what we did*). This could be an experiment, a reading, a discussion, or any other experience during which students figure out a key idea. The column designated for *what we observed* supports the results of a group discussion around what the students in the class agree they all observed or saw during the experience. The column indicating *what we figured out* is the space in which we record what the class decided in terms of coming to consensus on the disciplinary core idea(s) that they now understand as a result of this sensemaking/learning experience. In the final column, we record *how the core idea(s) relates to the phenomenon or problem under study.* It is a space in which students can literally and visually relate the core idea(s) back to the overall anchoring event they have been trying to figure out or solve. In this way, summary tables serve as a place to organize the collective thinking of the classroom community and display their ideas publicly. Similar to the DQB, the summary table is not just about displaying this publicly (although important), but it also represents a key opportunity to synthesize learning during the process of sensemaking. This is a powerful step in the group's learning process because in using this tool, all students need to agree on what they did or did not figure out based on their experience of collective sensemaking. Therefore, they need to be able to clearly articulate thoughts and ideas on the summary table. The additional benefit from use of summary tables is that these public representations of thinking can now be used by all students to provide context during the unit, and the information contained in the table can be used to remind students where they are in their learning. This is helpful for students who need additional support, as well as for those who might have been absent during a lesson, or even students who have been off-task and need a little nudge back in the right direction.

Beyond the science notebook, the DBQ board, and the summary table, there are many other tools that support group work in sensemaking lessons. These include general pedagogical strategies such as *think-pair-share* (a cooperative learning strategy) and *jigsaws* (another cooperative learning method). In addition, specific activity structures like gallery walks; group models; and three stay–one stray contribute to the sensemaking classroom. While the goal of this chapter is not to describe these additional teaching tools in detail, note that there are many online resources that provide more information. For example, see Ambitious Science Teaching (https://ambitiousscienceteaching.org/) and STEM Teaching Tools (https://stemteachingtools.org/). Overall, I want to highlight the fact that there is not one single way to design and implement a sensemaking lesson emphasizing group work. In fact, a suite of resources is available to use and customize to meet the needs of your students and your classroom.

SENSEMAKING MEANS THAT IDEAS CHANGE OVER TIME

Learning is a journey, not a destination. As educators, we've all seen how students build their understanding bit by bit, like putting together a complex puzzle. It's fascinating to watch as students gather pieces of information and gradually assemble the pieces into more sophisticated explanations, models, or designs. In my classroom, modeling is a tool that helps both to ground what we currently know and to identify gaps in our thinking (both individually and collectively). These include initial or partial models, rough sketches, or working hypotheses/conjectures that students create and that are not meant to be perfect—far from it! Instead, models function as snapshots of students' current thinking, ready to be refined and expanded as we dive deeper into a topic. I like to think of all models as working tools that students use as they engage in the process of knowing—as opposed to creating polished summative artifacts that are only shared at the end of a unit. Real learning happens when we revisit our models throughout a unit. Each lesson becomes an opportunity to test, critique, and refine our collective understanding. Of course, this approach requires careful planning. As teachers, we need to create multiple touchpoints for students to synthesize their learning across lessons. I find it helpful to start each class with a quick discussion: "What did we figure out yesterday, and based on that, where should we go next?" This not only reinforces prior learning, but also helps students see the bigger picture of how scientific ideas build on each other.

Sometimes the need for revision comes from unexpected places. Maybe we introduce a new phenomenon that challenges our current model. Or perhaps a heated classroom discussion reveals a gap in our understanding. In a sensemaking classroom, these moments are golden! They show students that science is a dynamic process, always open to new evidence and insights. By embracing this incremental approach, we create classrooms that buzz with authentic scientific inquiry. Students become active participants in building knowledge, not just passive recipients of information.

SENSEMAKING BUILDS INCREMENTALLY OVER TIME

As a STEM educator, I've always been fascinated by how students build their understanding of a phenomenon over time. When sensemaking is happening, it's like watching an extremely engaging new TV series where every episode helps you to uncover something new, and you cannot wait to move on to the next one. Clearly, sensemaking units are not just about teaching isolated facts. They are about weaving a rich tapestry of scientific understanding. Each unit becomes a new chapter in our ongoing science story, building on what we have learned before. Of course, making these connections explicit is crucial. I often start a new unit by asking, "Remember when we learned about ______? How do you think that might relate to what we're studying now?" In that way, every single thing we focus on as a classroom community is centered around what we currently know, and how we can leverage that knowledge to help us figure out new core ideas.

The sensemaking approach isn't just about content knowledge, though. It's also about helping K–12 students become more flexible thinkers. As they progress through the grades, they're not just learning more facts. With sensemaking, they are developing a more sophisticated understanding of how to use scientific and engineering practices and crosscutting concepts. I love watching this growth unfold. A student who may have struggled to identify patterns in simple datasets in elementary school might be better able to use their prior experiences with crosscutting concepts to confidently analyze complex ecosystem interactions in middle school. It's not that the earlier work wasn't important or was not done correctly—it laid the groundwork for enabling use of these more advanced skills later.

Ultimately, by taking a coherent sensemaking, building-block approach in our instructional practices, we're doing more than just teaching STEM concepts. We're helping students see the interconnectedness of the natural world. We're equipping them with the tools needed to make sense of new phenomena and to solve new problems they encounter both in and out of the classroom. And while it should go without saying, this type of learning is not just for students who primarily see themselves as STEM learners. All students need meaningful, relatable, and engaging experiences in STEM to open up the world of possibilities for them in educational settings and beyond.

EQUITABLE SENSEMAKING

Creating classrooms that truly support all learners is no small feat, especially given the incredible diversity of our students and the deep-seated inequities in our society. But it's a challenge we simply must tackle head-on in everything we do. When I think about equitable instruction, I'm reminded of a conversation I had with a colleague who works in a large urban school district. She told me, "TJ, it's not about creating additional activities for each student, but rather it's about reimagining how we teach so that every single student can thrive." And she's absolutely right. The materials we create for teachers need to be both robust and flexible. They should guide educators toward implementing practices that work for everyone, while still being adaptable to local contexts. Such equity practices aren't some add-on or afterthought—they're the beating heart of effective teaching and learning for all students.

At its core, the sensemaking approach to designing and teaching science lessons to K–12 students is about justice. It's about recognizing and valuing the multiple ways of knowing and being that our students bring to the classroom. Our materials should actively promote the rightful presence of all students by considering the dimensions of race, class, gender, Indigenous rights, immigration history, sexual orientation, and abilities. The list goes on. We need to take a hard look at the inequitable systems and assumptions baked into many of our educational institutions. By disrupting these, we open up new possibilities for learning that truly center on the lives and experiences of students from nondominant communities. Thus, the challenge of designing lessons for equity isn't just about the end product—it's about the process, too. We need diverse design teams that reflect the experiences and backgrounds of the students who'll be using these materials. And let's be clear, this is not about tokenism. We need meaningful integration of equity-focused expertise throughout the entire curriculum planning and lesson development processes. I'm reminded of what the *Framework for K-12 Science Education* tells us—all learning is a cultural accomplishment. By embracing this idea and building it into our instructional design, we can create materials that don't just pay lip service to equity, but truly embody it. The work of planning and teaching from the sensemaking perspective isn't easy, and we won't get it perfectly on the first try. That said, by committing to the sensemaking approach, we're taking crucial steps toward creating science classrooms in which every student finds learning meaningful, engages deeply, and thrives as a learner. When we talk about STEM education, it's crucial to remember that science, technology, engineering, and mathematics aren't the domain of any one group. Throughout history, brilliant minds from all walks of life have pushed the boundaries of what we know and what we can do. Our instructional materials need to reflect this rich tapestry of contributions.

Upon reflection, I am reminded of many conversations that I have had with students at Boston University who expressed their frustration with the lack of representation in their K–12 textbooks. In one of my courses—STEM Learning, Action, and Impact—we have a discussion in which many students share comments like this one: "It's like these kids are supposed to believe that only a handful of white guys in lab coats ever did anything important in science." I love when the issue of representation comes up because although students are (sadly) still having these kinds of experiences, they are beginning to see the truth and know that this inequitable situation needs to change. I suggest that this change starts with our curriculum. Our materials should be bursting with diverse images and stories that show the true faces of STEM—past, present, and future. I am not just talking about ticking boxes here. This kind of change is about creating inclusive storylines and examples of scientists and discoverers, from all backgrounds, that will resonate with all of our students. Whether the subject is community projects, family pursuits, or tackling global challenges, we need to show the many ways STEM intersects with our lives and our world.

Moreover, when we highlight contributions to STEM from members of diverse communities, presentations cannot be superficial. As science educators, our accounts should be substantial, accurate, and respectful. And let's not fall into the trap of only showing diversity in historical contexts. We need to make visible the incredible STEM work happening right now in diverse communities. Integrating cultural and gender diversity into our lessons shouldn't feel forced. It should be woven seamlessly into the fabric of our materials, connecting subject matter with relevant sociocultural contexts. For instance, when we talk about scientists, we should be showcasing individuals of all identities. This involves more than simple representation—it's about expanding our students' sense of global community and social responsibility. It is also crucial that we avoid stereotyping or essentializing cultural groups. Cultures are dynamic and varied, and our materials should reflect that. We need to highlight different approaches to building knowledge and making sense of the world. Remember, our goal isn't just to teach content. It's to build relationships and help every student see that they have a place in any STEM endeavor. When we get this right, we're not just improving STEM education—we're opening doors to futures that some students might never have imagined for themselves.

When we talk about truly impactful instructional materials, we need to think beyond just the content and include a focus on making connections to our students' lives, their communities, and the world around them. This isn't just a "nice-to-have" proposition. Making such connections in the classroom is essential for meaningful learning to take hold. I remember visiting a classroom teacher who had her students take photos of everyday examples of Newton's first law in their own lives. Newton stated, "An object at rest remains at rest, or if in motion, remains in motion at a constant velocity unless acted on by a net external force." The energy in that room was palpable. Students were excitedly sharing pictures of skateboard tricks, sports they played on the weekends, and even their pets playing. Suddenly, this abstract concept became real and personal. But we can't stop here. Our instructional materials and lesson designs need to go further, tapping into the dynamic practices and concerns of our students' communities. Imagine a unit on environmental science that doesn't just cover the basics, but dives into local water quality issues or air pollution concerns. That's how we make science relevant and actionable.

Real relevance in schooling means broadening our perspectives on what counts as scientific knowledge. Take, for example, a unit on plant biology. Rather than just planting a seed in a cup or putting a lima bean in a plastic bag and taping it to the window, the students learn about the planting method colloquially called the "Three Sisters." This method is a traditional Native American agricultural practice that involves interplanting corn, beans, and squash in close proximity, creating a mutually beneficial relationship among the plants. Corn provides structural support for climbing beans, and beans fix nitrogen in the soil, which benefits all three crops. Squash, with its large leaves, acts as living mulch, preserving soil moisture and suppressing weed growth. From a scientific perspective, the Three Sisters method demonstrates several key ecological principles: (a) symbiotic relationships, (b) resource partitioning, (c) nitrogen fixation, (d) microclimate modification, (e) climate resilience, (f) ethnobotany, and (g) cultural ecology, among others.

By incorporating the Three Sisters planting method into their curriculum, science teachers can provide students with a hands-on, integrative learning experience. This approach not only teaches specific scientific concepts, but also demonstrates the value of traditional ecological knowledge in understanding complex environmental systems. It offers a platform to discuss sustainable agricultural practices, biodiversity, and the importance of cultural perspectives in scientific understanding. Additionally, this method can be adapted for classroom or school garden projects, enabling students to observe and measure plant growth, soil health, and ecological interactions over time. Such practical applications can enhance students' engagement with scientific concepts and methods, while also fostering an appreciation for sustainable practices and cultural diversity in science.

As another example, I once observed a unit in which the middle school teacher's passion for gardening became central to bringing the curriculum to life and making it truly meaningful for students. Rather than just plant random things in a small outdoor raised-bed garden, the class decided to engineer the best garden for their entire school community. While this unit was one that I had helped to design, I admittedly was only a year or two into my understanding of the NGSS and still had a lot to learn about the NGSS principles and standards. That said, a highlight of creating this unit was that we had received local funding to provide each school in Hartford, Connecticut (a large urban district) with free kits of materials for use in these units. The particular unit I had co-created even included the materials to make a raised garden—down to the interlocking wood walls, wheelbarrows, soil, seeds, and shovels. Further, rather than the teachers planning and planting the garden themselves, they charged each class with engineering the "best" garden for their school. Keeping this engineering challenge broad in scope enabled the students to draw on their own knowledge and experiences. As such, they were able to create a design that met a number of criteria and constraints that they themselves determined. Then the class working on this project polled other students from across different grades and found out that the students wanted to grow things they could eat. Chips with salsa was by far the winning idea! They researched what goes into salsa by looking at food labels and family recipes. Then the class spent time thinking about the needs of the plants they would use based on their own garden's location. They planted peppers, tomatoes, and herbs and carefully measured their growth. They also used their math skills to estimate when they would have enough to harvest, with the idea of planning a salsa party for the whole school using the fruits of their own labor!

We also need to think about how we are nurturing our students' interests beyond the classroom. Every unit should include time for students to explore and reflect on their developing passions. Whether it's talking to an expert or diving into research on community issues, we're not just teaching STEM topics. We're cultivating lifelong learners and engaged citizens. To do this, we cannot forget the importance of honoring local histories and cultures, especially those of Indigenous communities. This isn't about tokenism or a one-off lesson. It's about weaving these perspectives throughout our curriculum, shaping how we frame ideas, how we approach pedagogy, and even the language we use in instruction.

SENSEMAKING: STUDENTS AS EXPERTS

When we talk about creating truly equitable learning environments, we need to shift our perspectives on student contributions. Students know when they are just "playing school" and when they are working on projects that capture their interest. This is why students thrive during project-based and inquiry-based lessons. So, how do we weave these components into all of the lessons we do? It starts with how we position students as experts who have an active role in their own sensemaking—as opposed to positioning students as passive receivers of the content delivered by the adult in the room. This can take many forms. Sometimes, positioning the student as an expert involves taking a student's seemingly off-topic contribution seriously, along with using *talk moves* (i.e., "say more") to better learn about how and why the student sees the two ideas as related. Positioning students as experts could also mean creating collaborative groups in which each student has an opportunity to be the expert in their group. While one student may not be as interested in the topic at hand, that student might shine when it comes to figuring out how to draw their thinking in a scientific model. Or a student who might not yet see themself as a STEM student may become really interested in creating a stop-motion–style video of how mitosis happens, in this way shining as a science communicator and developing an identity as more of an expert than they once thought.

Overall, we've got to broaden our view of what counts as competency. It's not just about mastering English or fitting into a narrow image of what a STEM person looks like. Instruction should celebrate the diverse ways students make sense of the world around them. A classroom that centers sensemaking helps teachers notice and leverage the unique ways students contribute to class discussions. In using the sensemaking approach—whether it's through speaking, writing, gestures, or diagrams—every form of expression is valued, contributing to the group's overall understanding.

SENSEMAKING AND MULTILINGUAL LEARNERS

When we talk about supporting multilingual learners, we're really talking about creating a classroom in which every student's linguistic and cultural background is seen as an asset, not a hurdle. It's about shifting our perspective from "How can we help these students catch up?" to "How can we tap into the incredible linguistic resources these students bring to the table?" I remember visiting a classroom where the teacher had students share their understanding of how plants get bigger (i.e., photosynthesis), using any language or mode of expression with which they felt comfortable. The result was a beautiful tapestry of explanations—some in English, some in Spanish, some through diagrams, and even one student who used a YouTube video and toggled on the captions. That's the kind of rich, multimodal learning environment we should be striving for.

Before moving too far, I want to highlight that my work with multilingual learners is heavily influenced by experts in this area, including Okhee Lee, Enrique Suárez, Scott Grapin, and María González-Howard. When we refer to these students as "emerging multilingual learners," we're not just playing with words. There are real concerns involved with the various terms people use to identify this category of students. Explaining why he uses the term "multilingual learners" rather than the term "English language learners," Grapin (2024) stated,

> I use the term "multilingual learner" broadly to encompass students bureaucratically labeled as English learners (ELs) by their schools as well as students who use multiple named languages and have faced persistent marginalization at the intersection of language with other identity markers (e.g., race). I use the term "multilingual learner education" broadly to encompass multiple disciplinary areas (e.g., bilingual education, TESOL, applied linguistics) that have contributed to research, policy, and practice related to the education of multilingual learners. While acknowledging the limitations of terminology that draws boundaries around language practices and disciplines that examine those practices, I use these terms in the spirit of unifying the efforts of a range of vested partners committed to the pursuit of educational equity for multilingual learners. (p. 168)

The distinction Grapin et al. (2022) described regarding his preference for the term "multilingual learner" builds on a report titled *Call to Action for Science Education* (National Academies of Sciences, Engineering, and Medicine [NASEM], 2021). This report noted that many teachers feel underprepared for teaching an increasingly diverse K–12 student population. It highlights the need for teachers to become "fluent in the subject matter they teach and fluent in the pedagogy of effective science instruction, including how to promote the success of culturally and linguistically diverse students" (NASEM, 2021, p. 23).

In addition, a 2022 National Science Teaching Association blog post from Suárez, Grapin, and González-Howard (which I frequently revisit; NSTA, n.d.) addresses their insights regarding ways to support science teachers in teaching multilingual learners. The bullet points below include three strategies and examples from their own work.

1. *Science teachers can adopt an asset-based view of multilingual learners in science classrooms.*

   Science teachers should recognize the strengths multilingual learners bring to the classroom, while challenging any unconscious deficit views. González-Howard shared that she has her preservice teachers design and conduct interactive science interviews with multilingual learners in which students explain their current understandings of a natural phenomenon. She shared that these interviews help her preservice teachers recognize the brilliance in students' ideas and their wide-ranging ways of expressing what they know.
2. *Science teachers can develop expansive understandings of language in science learning.*

   In science education for multilingual learners, teachers should critically examine their assumptions about language and its role in scientific practice. Understanding how different theories of language influence instructional approaches is crucial. Equally important is recognizing and valuing the multiple modalities through which students can represent and communicate scientific ideas, beyond traditional verbal or written expression. Grapin explained that the students in his science methods class examine different theories of language (e.g., language as structure vs. meaning-making) and reflect on how these components of language use impact student learning. He also has his students analyze student models in multiple modalities (e.g., drawings, symbols, language) so as to recognize the value of multiple ways of representing science ideas.
3. *Science teachers can develop a repertoire of practices for engaging multilingual learners in science learning.*

> Science teachers working with multilingual learners should integrate phenomena-driven learning with broader understandings of language in science. This approach involves recognizing and valuing the diverse communicative resources students use to make sense of a scientific phenomenon. By reflecting on their own sensemaking processes and observing multilingual students, teachers can appreciate the many ways to communicate scientific ideas beyond traditional academic English. Suárez (NSTA, n.d.) shared that in his professional development sessions, he asks teachers to figure out a phenomenon while paying special attention to the communicative resources they use to do so, giving the example of using onomatopoeia to illustrate the pitch of different sounds. He then has his PD participants view videos of students using similar methods to make meaning in science, highlighting the many ways that students can engage in sensemaking beyond the typically privileged "academic" English.

I appreciate this blog post because it helps me understand translanguaging in practice. Simply put, translanguaging acknowledges that people have one integrated system of language that includes the features of all the languages they know (Cioè-Peña & Snell, 2015). It is an approach that recognizes and leverages the full linguistic repertoire of multilingual learners in STEM classrooms. Unlike traditional views that treat languages as separate systems, translanguaging sees a multilingual person's language use as an integrated toolbox, ready to be used flexibly as needed.

In STEM classrooms, translanguaging supports student sensemaking by enabling learners to access their full linguistic resources when grappling with complex concepts. This promotes more equitable participation and can lead to deeper understanding. Translanguaging also supports metacognitive skills as students navigate between languages to explain STEM ideas. Implementing translanguaging in sensemaking classrooms might involve allowing students to brainstorm in their home languages, use bilingual resources, or present findings using multiple languages. This approach enhances conceptual understanding and also validates students' cultural and linguistic identities, potentially boosting multilanguage learners' engagement in STEM subjects. By embracing translanguaging, STEM educators can create more inclusive and effective learning environments, recognizing that the ultimate goal is STEM proficiency, not just English proficiency.

SENSEMAKING IN ELEMENTARY CLASSROOMS

In elementary classrooms (PK–5), sensemaking can take many forms, tailored to the developmental stages and experiences of young learners. According to the Framework and research on play and experiential learning, it is essential for younger students to engage with phenomena and problems that are tangible and relatable to them. For students in grades K–2, this means focusing on visible phenomena that they are likely to encounter in their everyday lives or within the classroom environment. These early experiences with sensemaking help build a foundation for scientific inquiry by enabling young learners to explore and question the world around them.

As students progress into the upper elementary grades (3–5), their sensemaking activities should evolve to include more complex and less visible phenomena. This shift encourages students to delve more deeply into macroscopic processes and systems, fostering a more comprehensive understanding of scientific concepts. At this stage, students are ready to explore topics that require them to think abstractly and understand the underlying mechanisms that drive observable events. For instance, they might investigate the water cycle, photosynthesis, or the interconnectedness of ecosystems, in this way gradually building a more nuanced appreciation for the complexities of the natural world.

By integrating these age-appropriate sensemaking experiences into the curriculum, science educators can create dynamic and engaging learning environments that nurture students' innate curiosity and support their cognitive development. In the early grades, focusing on visible and familiar phenomena helps build confidence and interest in science. As students grow older, the gradual introduction of more complex concepts challenges them to think critically and develop a deeper understanding of the natural world. Through this progression, sensemaking becomes a powerful tool for fostering a lifelong love of learning and scientific exploration.

How to Create Experiences That Foreground Sensemaking in Grades Pre-K-5

Anyone who has been around very young children knows that they are constantly making sense of their world. The baby who pushes a Cheerio off of the high chair table is running a little experiment of their own. What happens? Does it float up? Fly off 20 feet? Hmmm . . . it seems to fall down to the ground. Does anyone else see this? Do they come over? Does it draw the attention of the family dog, who might come over for a free treat? While this scenario isn't what most would call a science experiment, it does a lot for that young child in terms of gathering information about the world around them. Many sensemaking experiences in the elementary grades start in a similar way—very simple, but with the potential for students to learn a ton about how the world works.

For example, I was in a kindergarten classroom at a K–8 STEM magnet school located in the Northeast, and we were beginning a unit on Motion and Stability across all three classes: Forces and Interactions (NGSS standard: physical science #2 [PS2]). Rather than just begin the unit by explaining what forces and interactions are, we gave each student a simple version of a science notebook and told them we were going to start with a challenge. We instructed students that we would be going on a scavenger-style hunt around the classroom to see what could be pushed and what could be pulled. Very soon after giving this instruction, we reminded the young students that they were to push items in the classroom and not *people* in the classroom! The goal was to find a way to capture what kinds of things in the classroom could be pushed, pulled, or neither. Students used their science notebooks in various ways, as the goal was to introduce the notebooks as a tool, not for these young students to fill them with their writing. Some students immediately folded their notebooks under their arms and went around to push and pull on things. Others took a moment to look at their notebooks and ask the teachers how to write about something like this. We suggested that they could draw the item and then have two columns: one for push and one for pull. They could put a check mark in one of the columns after they tried pushing and/or pulling an item. Some students were more free-form than others with how they wanted to record what they were doing. Yet, when we prompted them to share what they found from their explorations, they could easily recount what they discovered after experimenting with each object.

Taking a step back—with a fly-on-the-wall perspective—I could clearly see evidence of rich sensemaking in this kindergarten classroom. Students were walking around and testing items (e.g., tables, rulers, shelves, walls, pencils, books) and writing about what they were finding. But increasingly, students were honing in on the kinds of things they were testing. Initially, there was a bit of an *everything goes* strategy whereby anything they saw was open to being pushed or pulled, with full attention on the result. Over time, the children focused on items that were smaller, pushing and pulling them with less surprise and then calmly adding to their notebooks.

When they approached some items that were larger (potentially harder to push/pull), the children placed their notebooks down somewhere nearby, as they anticipated needing extra effort to test the larger items. When we asked them why they thought it would require a bigger push or pull, they shared that other items of similar size (referencing their drawings in their notebooks) took more effort, so they thought that this item would require more strength, too.

This kindergarten class experience serves as an example of very young children engaging in a rich sensemaking investigation, carefully planned by the teachers, that set students up to immerse themselves in seeing their world through a new lens. The learning experience went from normal classroom *stuff* into a new world of possibilities where everything could be tested and captured, as evidence in their notebooks. The whole-class discussion that followed took the form of a consensus discussion, that is, a discussion during which students were facilitated by the teacher to try and see what they agreed upon based on their investigation. Initially, students were just sharing what they did. They pointed out things they tested and what they found. We reminded students of the established classroom norm—using a hand gesture, with thumb pointing to them and pinky finger extended toward who was speaking (other fingers down)—to signal that they agreed with the student who was talking. We also reminded them to share if they found something different; that our goal was not *correct* answers at this point, but rather sharing all of our evidence publicly. Students quickly decided that we should make a table like some students did. So as a class, we created a table that had a list of items from around the room and tally marks for evidence of pushes or pulls. Students quickly started seeing patterns (CCC) and shared more ideas like, "Things on tables are easy to push and pull, things on the floor are harder to move." Over the course of the discussion, more questions and ideas emerged. For instance, the idea that some things simply do not move, as well as the idea that each student might be pushing or pulling with more force, started to come up more and more. We shared that maybe our tally marks were not sufficient to capture what we were doing, that we might need to do something else to figure out how the *size* of the push or pull impacts whether or not something moves. This was a key goal of our discussion because while we were focused on science in this lesson (i.e., pushes and pulls can have different strengths and directions; K-PS2-1, K-PS2-2), our next lesson would be a mathematics lesson, and the push-pull experience was naturally leading students to want to move in the direction of being able to quantify the size of these changes in strength. We then created a classroom list of potential items to test (i.e., things that students thought to be heavy), with the discussion focusing on one particular set of paperweight-style bookends because they were a clear winner in terms of representing an item on which we were not all in agreement.

After we decided to focus on this heavy set of objects, students came up with all sorts of ways to test them. Some said we could measure how heavy they were. Others said we could measure the bookends with the rulers and just see which was bigger. We asked table teams to come up with a plan for our investigation, an instruction indicating a clear use of science practice (i.e., With guidance, plan and conduct an investigation in collaboration with peers; K-PS2-1). At the same time, we encouraged students to think about what kinds of data (numbers) we could collect that would convince us that an item needed more or less of a push or pull.

Mathematics Standards in This Lesson

MP.2: Reason abstractly and quantitatively.
K.MD.A.1: Describe measurable attributes of objects, such as length or weight. Describe several measurable attributes of a single object.
K.MD.A.2: Directly compare two objects with a measurable attribute in common, to see which object has "more of"/"less of" the attribute, and describe the difference.
(Common Core State Standards Initiative, 2010)

The way in which we integrated math into the lesson was not in the context of saying, "Okay, now we are going to do a math lesson." Rather, the use of math was motivated from a need to figure out more attributes about what we were already interested in investigating: kindergarten classroom items that needed to be either pushed or pulled. Students worked together to come up with their plans for investigating items' sizes and weights. Ultimately, we spent a whole class session exploring and coming up with ways to measure a huge number of attributes about various objects and how this additional information might impact the size of the push or pull. In addition, the class decided that the larger students did bigger pushes, and the smaller students did smaller pushes. Thus, as a classroom community, the students agreed that they would use the sizes of students to "accurately" control for the amount of push or pull.

THE ROLE OF TECHNOLOGY AND ENGINEERING IN SENSEMAKING FOR YOUNG CHILDREN

Before shifting to some practical suggestions for designing sensemaking units/lessons with elementary students, I think that it is worth highlighting how this same group of kindergarten students used technology to assist them in their sensemaking. I am staying with my focus on this kindergarten class because I feel it is important to show that very young children can engage in rich sensemaking experiences, and that we can often raise the bar higher than we think in what we ask of our students (with appropriate supports). The following example certainly highlights this proposition as it demonstrates how students worked together to code their own robots—a task that the class decided would be an effective way to collect data during severe weather!

We now drop in on this kindergarten class near the end of their STEM unit on weather. They had been focused on Physical Science and Earth Science Standards, learning all about types of weather and tracking weather conditions by using simple observations and a daily weather chart. With students seated on a rug, we began the session with a whole-class discussion. We started by recapping the kinds of weather that we had seen while going outside to observe and record the daily weather. I then shared a video that was a compilation of severe weather events that can be seen in the United States (e.g., tornadoes, derechos, hail, hurricanes). The students were immediately excited, exclaiming that this kind of weather is scary and that they could not believe what they were seeing. We then asked, "Which of you would be willing to go outside to tell us about the weather on a day like the ones we are seeing in the videos?" In each of the three kindergarten classes, there was always at least one student who bravely said they would volunteer (getting a reaction from their classmates), only to later say they were joking after we replayed the videos a second time. After viewing the videos a second time, we asked students to "turn-and-talk" with a neighbor to try to come up with ways to get weather information during times of severe weather like we were seeing.

When we regrouped, we asked students to share their ideas. Students had all sorts of ideas, like setting up phones to record their observations or putting buckets outside to capture the rain. Interestingly, in each class the topic of putting a robot outside came up. When this happened, we asked everyone to think about the possibility of using a robot, and if that would help us in this situation. We also shared some pictures of a vehicle that looked well engineered to go into storms, introducing them to the kinds of storm-chaser vehicles that currently exist. The students agreed that storm-chaser vehicles represented a productive direction to pursue. We then shared that, luckily, we had some robots in our school that we could use as our own version of storm-chasing vehicles to test the idea!

In fact, we had a set of Kibo robots at the school that had been purchased for use by the STEM Coordinator and Technology teacher. The Kibo robots were from KinderLab Robotics, a company that emerged from Dr. Marina Umaschi Bers's robotics research on educational technology in elementary classrooms. Kibo robots are about the size of a small lunchbox, with wheels on the sides and ports for students to attach sensors at the top. The screen-free robot has a scanner at the front, similar to a scanner at a grocery store, that can be used to scan small wooden blocks labeled with words indicating a motion (e.g., forward, backward, turn left/right, spin). We introduced the robots and showed students how to scan the blocks, giving them one class to explore the robots further. Next, seated as table teams, each team of students had a large, laminated sheet of construction paper that served as their home base. We instructed students to figure out a way to get the robot to drive off of the base, turn around, and get back to the paper in one fluid program. One quick note: We did not begin the lesson by explaining *how* to do this, but left it open for students to solve this tech/engineering challenge. Within about 20 minutes, every student had figured out how to do this and were already moving on to more advanced challenges that they had spontaneously invented in their groups. We regrouped on the rug to discuss what was challenging and why. Each group/team shared their successful plan by demonstrating what their robot was programmed to do.

During our next class, we again started on the rug with one of the Kibo robots in hand. The students all agreed that we could program the Kibo robots to follow a path of our choosing. The challenge now was to figure out how we could collect data on the weather while the robot was outside. While we had not yet introduced examples of any weather tools, students had investigated wind, rain, sun, and other types of weather. We reminded them that we could not fit inside the Kibo like the people in the storm-chasing vehicles. Using a provided platform that attaches to the top of the Kibo robot, we gave students a paper circle that had been pre-cut to fit on top of the platform. We then instructed the students to design their own weather tools that could collect weather data. Back at their tables, students discussed different things they could draw to represent ways of collecting weather information. While I was visiting tables, I would ask what they were working on in order to get students to go public with their ideas. Most table groups came up with the idea of drawing a fan (of sorts) to measure the wind, along with some sort of bucket or cup to measure the rain. I'll never forget the one group that had drawn a fan and an upside-down top hat. When I asked what the top hat was for, one student said "Ummm, you look in the hat to see how much rain is in there!" This exchange clearly pointed out that I was not thinking outside of the box like these brilliant students. Some of the drawings included VERY creative and innovative ideas (which do not exist *yet*), so we asked these groups to explain their reasoning for how their new ideas would work. It was important to us to not limit students' creativity, while also pressing for their reasoning and problem-solving skills to emerge.

In our next class, we introduced the final challenge: navigating their storm-chasing robot into the storm! I happened to have a large donation of fluffy white cotton at my disposal and, with it, built a storm on a pole that looked like a huge cloud with blue streamers hanging down to the floor. I positioned my storm creation in the center of the room. Note that before class, we had arranged each team's home base in a place around the room—with the classroom tables and chairs in the direct path of making it to the storm. Very quickly, the students were excited to get to work and start programming the robots to solve this engineering challenge. Because the blocks with bar codes were in limited supply, we photocopied the blocks and gave students a whole set of small squares with the movements (forward, backward, etc.) on them. We also gave them a sentence strip that we had folded into a holder of sorts. With this, they could slide the cards into the holder and easily reorder them, keeping the cards from flying all over the place. In each group, some students were focused on the cards, trying to guess how many of each movement they would need to solve the challenge. Other students began playing the role of the robot, taking steps forward and left around objects, and then back toward the storm, measuring each step so as to bring back numbers to their group.

A key component of engineering and technology lessons that involve sensemaking is the focus on careful planning rather than trial-and-error–type experiences. Instead of letting students just go for it and try their Kibo robot's program each step of the way, students needed to share their plan with the teacher and explain, with sound reasoning, why they thought their plan would effectively solve the challenge at hand. When the student groups felt that they were ready, they presented their series of steps (laid out in the folded sentence strips) for review. We asked, "How do you think your robot will do with this challenge?" The students responded by explaining their programs, describing each step in the process. Every group had one student who served as the robot, with that student making the moves as they were read. We observed that each student who was playing the role of the robot had different interpretations of what the robot would do for one step and one move left. They would find out what the robots actually did when they ran their program using the robots for the first time. However, if a group had a plan and all group members were in agreement about how it would work, then we approved their plan, and they were good to move forward and test.

When each group was ready, we called the class together to view each team's first attempt to reach the storm. Because they were so far away from the storm, each group ran into some sort of challenge when they hit GO on their Kibo. After the run-through, we asked each team to share what they thought about their first run and their initial ideas about how to improve their next trial. They shared things like "We need more forwards" or "The turn does not move far, so we need to turn and then go forward." After we heard from each group, we asked the rest of the class to comment on the current group's ideas for revising their plan. We did this purposefully to reinforce that this challenge was a classroom challenge, and everyone needed to support each other in reaching the storm. Each group shared what they thought about the potential revisions and offered insights from their own experimentation with the Kibo. By the end of one class period, each group had made it to the storm. They also figured out that they needed to reverse the program to get the robot back to them with the "weather data" it collected from the storm.

It is important to highlight that this unit was a truly integrated STEM unit that included science, technology, engineering, and math skills and concepts. While the Framework and NGSS include technology standards, it is worth calling out the International Society for Technology in Education standards (ISTE), as our instructional unit on weather drew heavily on the use of technology as a driver for student learning in math, science, and engineering. You can view the full ISTE standards at https://iste.org/standards.

The final experience in this kindergarten weather unit was focused on sensemaking as it related to STEM careers. It is important to help students see that the skills and knowledge they are gaining in school can be applied to a number of different careers that they might wish to pursue later in life. For this particular unit, we reached out to the local Channel 3 affiliate in our area and asked them to bring a meteorologist and their weather tracker vehicle for a classroom demonstration of how they work. Many television stations have a touch-the-truck–style initiative that involves bringing the weather tracker to visit schools and sharing their work experiences with weather. To clarify, the weather tracker is a vehicle that these stations use to collect weather data from different towns in their areas and broadcast the information live back to the news station.

As soon as the weather tracker arrived, students were yelling and pointing at all the weather tools it had on the roof. "That fan looks like the one on our Kibo!" "Look at all the weather tools!" The students quickly made the connection that this mobile weather station was just like their Kibo outfitted with the weather tools that they had designed. The meteorologist gathered each class (one at a time) around the weather tracker and explained what the weather tools were and how they collected weather data. He also invited the students inside the tracker to see all the live data, showing them how he uses the data to make predictions about what the weather will be in the future. He also told students what it is like to have to use science and communication skills to help him make sense of the weather in ways that the public can use to make informed decisions about their own lives during important weather events. This experience was truly transformative for these young students, and we continued to hear references to that day for years to come.

Practical Strategies for Sensemaking in Elementary Classrooms

While I could go on and on with examples of meaningful sensemaking lessons, including lessons and units for upper elementary students, the goal of this book is to provide practical examples of how you can incorporate sensemaking into your own classroom. The list below is not an exhaustive list, nor is it meant to be completely linear in how one should proceed through each item. Instead, the following list of recommendations is meant to provide a number of strategies that can be examined when planning for sensemaking units and lessons.

1. **Identify the targeted performance expectations:** Select the specific NGSS performance expectations your lesson will address.
2. **Integrate the three dimensions:** Ensure that your lesson incorporates relevant DCIs, SEPs, and CCCs.
3. **Start with a phenomenon or problem:** Introduce a real-world scientific phenomenon or engineering problem that sparks curiosity and requires investigation.
4. **Engage in science and engineering practices:** Design activities that allow students to actively engage in practices such as asking questions, developing models, planning investigations, analyzing data, or constructing explanations.
5. **Emphasize crosscutting concepts:** Highlight how CCCs like patterns, cause and effect, or systems and system models apply to the phenomenon or problem.

6. **Use driving questions:** Frame the lesson around questions that guide student inquiry and sensemaking.
7. **Promote evidence-based reasoning:** Encourage students to use evidence from their investigations to support claims and construct explanations.
8. **Facilitate scientific discourse:** Create opportunities for students to engage in scientific arguments based on evidence and critique of others' reasoning.
9. **Develop and use models:** Incorporate activities where students create, use, evaluate, and revise models to explain phenomena or solve problems.
10. **Connect to nature of science concepts:** Help students understand how scientific knowledge is developed and how science itself works.
11. **Integrate engineering design:** Where appropriate, include engineering challenges that allow students to apply scientific knowledge to solve problems.
12. **Expect and include opportunities for misconceptions to arise:** Make sure to plan for common misconceptions to emerge, and include evidence-based investigations that provide alternative explanations for these topics.
13. **Use formative assessment strategies:** Continuously assess student understanding and adjust instruction using techniques aligned with three-dimensional learning.
14. **Make connections across disciplines:** Highlight how the concepts relate to other scientific disciplines and real-world applications.
15. **Reflect on learning:** Guide students in reflecting on how their understanding has changed and what questions they still have.

SENSEMAKING IN SECONDARY CLASSROOMS

In secondary classrooms (6–12), sensemaking often presents unique challenges due to the time required to deeply explore phenomena and solve complex problems. This approach contrasts with traditional lectures and labs that prioritize content delivery and confirmatory experiments. However, prioritizing sensemaking in these grades is crucial, especially as middle school students begin to tackle microscopic phenomena such as atoms, molecules, and cells. In high school, the complexity of scientific concepts escalates as students delve into subatomic particles and explore intricate interactions within and among systems at various scales. Sensemaking remains a critical component of science education at this level, enabling students to build on their foundational knowledge and tackle advanced topics. These topics are inherently complex and require students to engage actively with the material, articulate their reasoning, and use science practices to gather and interpret evidence.

As students progress through middle school, their ability to explain and understand microscopic phenomena benefits significantly from a sensemaking approach. This further builds in high school with students moving toward more complex explanations, models, and arguments about the interconnectedness of the natural and designed world around them. This process involves students making their thinking public, discussing and debating ideas with peers, and continually refining their understanding based on new evidence. By leveraging science practices—such as developing models, carrying out investigations, and constructing explanations—students can navigate these challenging topics more effectively. This method not only helps in solidifying their understanding, but also prepares them for the increasingly complex scientific concepts they will encounter in high school.

Not the Way We Experienced Middle and High School

Traditional educational methods can often perpetuate notions such as *it worked for me* or, similarly, *we've always done it this way*. These excuses lead many of us who work in the field of education to ignore new understandings of how people learn. Thus, we continue to teach in a way that does not reach all of our students. More specifically, this approach to teaching continues to make STEM seem like an identity that is only for some. The shift from "learning about" to "figuring out" reframes learning from being a passive experience for students and, instead, situates students as being actively engaged in the process of exposing their own gaps in understanding, in turn working to fill those gaps through engagement in the science and engineering practices. In this section, I highlight a few key ways to shift the conventional design of secondary classroom experiences so that students can begin their shift toward being the ones doing the heavy lifting in leading the sensemaking process—all through expertly designed teacher-led experiences. I then share a series of practical strategies for sensemaking in secondary classrooms.

One clear example of this shift is to purposefully think about our habitual or old ways of starting or introducing units in secondary classrooms. These more traditional ways often began with a set of slides and definitions that students are meant to understand *before* starting the unit, or a demonstration of a lab that walks students step-by-step through exactly what they are to do, and what they should experimentally conclude from in their experiment. In contrast, starting units with sensemaking in the foreground calls for phenomena and problems to be at the center of our work in the classroom. Doing so becomes more complicated as the topics become more complex in the later years of K–12 education. Therefore, we often want to start with phenomena that equal this complexity and opt for things like climate change or a new species introduced into an ecosystem (e.g., wolves in Yellowstone)—in other words, phenomena that match the level of complexity. But it is important to remember that we are trying to provide our students with the skills to approach STEM topics like experts in these fields do. I know from working with STEM graduate students and professors in the field that they still approach these topics from a practical and step-by-step approach. We should help students do the same. When these topics are overwhelming, it is easy to want to give up because we are unsure of where to start, or sometimes, we are so perplexed by the topic that we do not see a path to understanding it all.

I suggest that rather than trying to tackle huge topics, we should help secondary students focus on making sense of important but tangible components of these larger systems. In this way, we can avoid making units over-the-top grandiose. For example, imagine beginning a high school unit on climate change and introducing students to the topic by starting with a scavenger hunt aimed at exploring local impacts on the environment. At this early stage in their understanding, it is likely that students will find evidence of things that both are—and are not—connected to the topic of climate change. In fact, such variable evidence would lead to a wonderfully diverse set of questions and ideas presented on a driving question board that could motivate our collective sensemaking on the topic. For instance, in a unit on climate change that we use for our NGSX unit on Oceans, Climate Science, and Big Data, we decided early on to focus on harmful algal blooms (HABs) as a first step into the topic. With this unit, it has been productive to start with HABs because we can see evidence of these blooms no matter where you live (ocean areas and lake areas), and they can be indicators of larger problems in the environment. These HABs, while seemingly small in scale compared to the vast topic of climate change, provide a tangible entry point for students to begin their explorations. By focusing on something they can observe and potentially even experience in their local environment, we create a bridge between the abstract concept of global climate change and its concrete, observable effects. This approach aligns with the "figuring out" model we discussed earlier. When students investigate HABs, they're not just passively receiving information about climate change. Instead, they're actively engaged in questioning why these blooms occur, what conditions facilitate their growth, and how they might be connected to larger environmental shifts. This process naturally leads to broader questions about water quality, temperature changes, and nutrient runoff—all of which are integral to understanding climate change.

To further illustrate, starting a climate change unit with HABs allows for a gradual buildup of complexity. As students become more comfortable with the concept and its immediate impacts, we can guide them to explore the interconnections between HABs and other environmental phenomena. This scaffolded approach helps prevent the overwhelming feeling that often accompanies diving headfirst into a topic as vast as climate change. It's crucial to remember that our goal isn't just to cover content, but also to develop students' scientific thinking and inquiry skills. By starting with a focused, observable phenomenon like HABs, we're not simplifying the science. Instead, we're making it more accessible and engaging. We're teaching students how to approach complex problems by breaking them down into manageable parts, a skill that will serve them well in any scientific endeavor. In practice, this might look like students collecting water samples from local bodies of water, analyzing the samples for algal content, and comparing their findings with historical data. Students might interview local environmental experts or reach out to local fishermen to ask about changes they have observed over time. These hands-on experiences make the learning more engaging, and they help students see themselves as active participants in scientific inquiry—not just passive recipients of information.

Authenticity is key in reshaping how high school students perceive and engage with sensemaking in STEM. In high school classrooms, authenticity means moving beyond textbook problems and contrived scenarios. It's about presenting students with real-world challenges that don't have neat, prepackaged solutions. For instance, instead of asking students to balance chemical equations on paper, we might challenge them to investigate the chemical composition of local water sources experiencing HABs and the impact of HABs on aquatic life. The shift from a theoretical to a practical orientation not only makes the learning more engaging, but it also helps students see the relevance of STEM in their everyday lives. Authentic sensemaking experiences also involve embracing uncertainty. In the real world, scientists don't always have all the answers, and neither should our students. By presenting them with open-ended problems, we're teaching them that it's okay not to know everything immediately. The process of figuring things out, while grappling with uncertainty, is where real learning happens. This approach mirrors the actual work of scientists and engineers, who often start with more questions than answers. Additionally, authenticity in STEM sensemaking means connecting with the community and industry. Bringing in local experts, collaborating with nearby research institutions, or partnering with community organizations on real projects can provide students with a window into how STEM is applied beyond the classroom walls. For example, students might work with local environmental groups to monitor air quality in their neighborhood, analyzing data and presenting findings to city officials. Such experiences not only deepen students' understanding of scientific concepts, these experiences also show students the significance of the impact that their STEM knowledge can have on their community.

Rethinking How We Do Labs

The Framework for K–12 Science Education has fundamentally challenged our traditional approach to laboratory work in science classrooms. It's time we take a hard look at how we conduct labs and ask ourselves, "Are we truly fostering scientific thinking, or are we simply going through the motions?" Traditionally, many of our lab experiences have been what we might call "cookbook" labs. Students follow a prescribed set of steps, often knowing the expected outcome in advance. While these labs might help reinforce concepts, they fall short in developing the critical thinking and problem-solving skills that are at the heart of scientific practice. The Framework pushes us to reimagine labs as opportunities for authentic scientific inquiry. Instead of handing students a procedure, we should be presenting them with phenomena or problems to investigate. For instance, rather than giving students a step-by-step guide to observe osmosis in plant cells, we might challenge them to figure out why cut flowers wilt faster in saltwater than in freshwater. This approach not only engages students more deeply, but it also mirrors the way real scientists approach unknowns in their work.

Additionally, the Framework emphasizes the importance of the science and engineering practices. Our labs should be spaces where students actively engage in these practices—asking questions, developing and using models, planning and carrying out investigations, analyzing and interpreting data, and constructing explanations. This means moving away from labs where the primary goal is to "get the right answer" and moving toward experiences through which the process of investigation is valued as much as the outcome.

We also need to reconsider the role of "lab reports" in our classes. While there's value in students documenting their work, the traditional lab report format often becomes a fill-in-the-blank exercise that doesn't reflect authentic scientific communication. Instead, we should explore diverse ways for students to share their findings—perhaps through scientific posters, peer presentations, or even mock grant proposals. These formats can better prepare students for how scientific ideas are actually communicated in the real world. Another key shift involves how we approach "failed" lab experiments. In many traditional labs, if students don't get the expected result, it's seen as a failure. But the Framework reminds us that unexpected results are often the most valuable in science. We should be designing lab experiences that encourage students to grapple with unexpected outcomes, question their methods, and revise their approaches. Such experiences teach resilience and provide students with a more authentic understanding of the scientific process.

The Framework also calls for a greater integration of technology in our lab work. This doesn't mean replacing hands-on experiences with simulations, but, rather, using technology to enhance our investigations. For example, students might use digital probes to collect data over extended periods, analyze large datasets to look for patterns, or use modeling software to test hypotheses that would be impractical to investigate physically. Ultimately, rethinking our approach to labs based on the Framework means shifting from a focus on verification to one of exploration and discovery. It's about creating spaces where students can engage in the messy, iterative process of science—where they can ask their own questions, design their own investigations, and construct their own understandings of the world around them. This approach aligns with what we know about how students learn best, and it better prepares them for the realities of scientific work in the 21st century.

Practical Strategies for Sensemaking in Secondary Classrooms

Luckily, there are more and more examples of high-quality instructional materials being released that foreground sensemaking. While I always enjoy reading about examples of these sensemaking lessons and units at the secondary level, the goal of this book is to provide practical recommendations regarding how you can incorporate sensemaking into your own classroom. The list below is not an exhaustive list, nor is it meant to be completely linear in how one should proceed through each item. Instead, the following list is meant to provide a number of strategies that can be examined when planning for sensemaking units and lessons for secondary classrooms.

1. **Focus on depth over breadth:** Design lessons that allow for deep exploration of core ideas rather than superficial coverage of many topics.

2. **Use anchoring phenomena:** Begin units with complex, puzzling phenomena that require multiple science concepts to explain fully.
3. **Develop student-led investigations:** Encourage students to design and conduct their own scientific investigations to answer questions about phenomena.
4. **Emphasize systems thinking:** Guide students in analyzing complex systems, including feedback loops, emergent properties, and system boundaries.
5. **Integrate mathematical and computational thinking:** Incorporate data analysis, modeling, and simulations to strengthen connections between science and mathematics.
6. **Engage in argument from evidence:** Provide opportunities for students to construct and critique scientific arguments based on evidence and reasoning.
7. **Address socio-scientific issues:** Explore the intersections of science, technology, society, and the environment through current, relevant topics.
8. **Implement engineering design challenges:** Incorporate more complex, multi-step engineering projects that require application of scientific principles.
9. **Use scientific modeling:** Engage students in developing, testing, and refining conceptual, mathematical, and computational models.
10. **Explore career connections:** Highlight how the science concepts and practices relate to various STEM careers and current research.
11. **Incorporate scientific literature:** Have students read, analyze, and discuss scientific papers and reports appropriate for their level.
12. **Emphasize nature of science:** Delve deeper into how scientific knowledge is constructed, validated, and revised over time.
13. **Foster interdisciplinary connections:** Design projects that integrate concepts from multiple scientific disciplines (e.g., biology, chemistry, physics).
14. **Utilize advanced technology:** Incorporate more sophisticated scientific tools and technologies in investigations and data analysis.
15. **Develop scientific communication skills:** Have students create scientific posters, give presentations, and write lab reports in styles similar to professional scientists.
16. **Explore ethical considerations:** Discuss ethical implications of scientific research and technological advancements.
17. **Implement project-based learning:** Design long-term projects that allow students to investigate real-world problems in their community or beyond.

18. **Foster metacognition:** Encourage students to reflect on their own thinking processes and how their understanding of scientific concepts evolves.
19. **Address misconceptions at a deeper level:** Tackle more complex misconceptions and help students reconcile scientific explanations with their intuitive understandings and new evidence.
20. **Prepare for college and career readiness:** Align activities with skills and knowledge needed for success in college-level science courses and STEM careers.

SENSEMAKING IN OUT-OF-CLASSROOM SETTINGS

Leveraging phenomena in out-of-classroom settings can greatly enhance students' sensemaking experiences, providing opportunities that go beyond the confines of traditional classrooms. Field trips to science centers, museums, and natural environments are often underutilized in terms of their potential to deepen students' understanding of scientific concepts. Too often, these excursions are treated as stand-alone events rather than integral components of a cohesive learning journey. However, with thoughtful integration, these out-of-school settings can become invaluable extensions of the classroom, offering opportunities to consider unique phenomena that spark curiosity and drive inquiry.

Drawing on my extensive experience in designing learning experiences and exhibits, I have found that the key to successful integration lies in intentional planning and alignment with classroom goals. For instance, a visit to a science museum can be more than a passive observational experience. By designing pre-visit activities that introduce relevant phenomena and post-visit projects that encourage students to apply their observations, educators can create a seamless connection between the field trip and the classroom curriculum. This approach reinforces the concepts learned in class and provides students with a richer, more engaging learning experience. In fact, science centers and museums are reservoirs of phenomena that are often inaccessible within the classroom. These venues offer interactive exhibits, live demonstrations, and real-world examples that bring abstract concepts to life. For example, a planetarium show can vividly illustrate astronomical phenomena that students have studied in textbooks by providing a three-dimensional perspective that enhances their spatial understanding. Similarly, a hands-on exhibit about the physics of roller coasters can help students grasp complex principles of motion and energy through direct manipulation and experimentation. By leveraging these resources, educators can help students make concrete connections between theory and real-world applications.

Field experiences in natural settings—such as parks, botanical gardens, and nature reserves—offer another dimension of learning that is particularly powerful for sensemaking. These environments provide opportunities for students to engage directly with the natural world, where they can observe ecosystems, collect data, and conduct experiments in situ. For instance, a trip to a coastal ecosystem enables students to investigate the impacts of human activity on marine life, linking their observations to broader environmental science concepts discussed in class. By structuring these experiences to include guided inquiry and reflection, educators can help students develop a deeper appreciation for the interconnectedness of natural systems and the relevance of scientific inquiry to everyday life. To truly integrate an out-of-classroom learning experience with the sensemaking process, it is crucial to adopt a holistic approach that considers the entire learning cycle. This involves careful planning to ensure that field trips are not isolated events but are, in fact, embedded within a larger instructional framework. Pre-visit activities should introduce key concepts and phenomena, setting the stage for what students will encounter in the out-of-classroom setting. During the visit, structured activities and guided observations can help students focus their attention and engage actively with exhibits or natural settings. Post-visit projects and discussions should then provide opportunities for students to synthesize their experiences, draw connections to classroom learning, and extend their investigations further.

By redefining how we utilize science centers, museums, and field experiences, we can transform educational outings into powerful tools for sensemaking. These settings offer unique opportunities for students to engage with phenomena that are not available within the typical classroom environment, enriching the learning process and fostering a deeper, more authentic understanding of science. Through careful integration and intentional design, educators can leverage out-of-classroom experiences in ways that will enhance students' sensemaking abilities and make science education more dynamic, relevant, and impactful.

PRACTICAL STRATEGIES FOR SENSEMAKING IN OUT-OF-SCHOOL SETTINGS

The kinds of learning experiences that students can have in out-of-school settings and natural environments are unlimited. An entire book showing how such diverse and robust experiences can support student learning could (and probably should) be written and published for education professionals. Nonetheless, the goal of this book is to provide practical examples of how you can incorporate sensemaking in your science class units and lessons. The list of recommendations below should provide a number of practical ways that out-of-school experiences—like field trips and after-school programs—can support student learning with sensemaking at the center.

1. **Leverage local phenomena:** Focus on scientific phenomena that are observable in the local environment or community.
2. **Design place-based investigations:** Create activities that allow learners to explore and investigate their immediate surroundings.
3. **Incorporate citizen science:** Engage participants in real scientific research projects that contribute to larger databases or studies.
4. **Use everyday materials:** Design experiments and activities using readily available materials to make science accessible outside formal settings.
5. **Connect to learners' interests:** Tailor activities to align with participants' hobbies, cultural backgrounds, or community issues.
6. **Emphasize hands-on experiences:** Prioritize tactile, experiential learning that may not be feasible in traditional classroom settings.
7. **Foster intergenerational learning:** Create opportunities for learners of different ages to collaborate and learn from one another.
8. **Integrate technology meaningfully:** Use mobile devices, apps, and online resources to enhance data collection, analysis, and sharing.
9. **Promote environmental stewardship:** Design activities that connect scientific understanding to local environmental issues and conservation efforts.
10. **Collaborate with community partners:** Work with local organizations, businesses, or experts to provide authentic contexts for learning.
11. **Emphasize crosscutting concepts:** Highlight how scientific ideas connect across different contexts and disciplines in everyday life.
12. **Encourage long-term projects:** Design activities that can be revisited over time, allowing for extended observation and data collection.

13. **Incorporate engineering challenges:** Create design challenges that address real community needs or problems.
14. **Foster science communication skills:** Encourage learners to share their findings with the community through presentations, exhibits, or social media.
15. **Explore STEM careers informally:** Introduce learners to diverse STEM professionals and career paths through informal interactions or mentorship.
16. **Utilize informal assessment strategies:** Implement creative ways to gauge understanding, such as science journals, photo documentation, or peer discussions.
17. **Promote family engagement:** Design activities that encourage family participation and learning together.
18. **Incorporate artistic expression:** Integrate art and creativity into science activities to appeal to diverse learners and multiple intelligences.
19. **Address social justice issues:** Explore how science and engineering can address issues of equity and social justice in the community.
20. **Cultivate science identity:** Help learners see themselves as capable of doing science and engineering in their everyday lives.

EMBRACING SENSEMAKING AS A PATH FORWARD

As we wrap up this short dive into designing sensemaking experiences in STEM education, it should be clear that we're not just talking about a new teaching technique. In fact, we are discussing a fundamental shift in how we approach teaching science and how we foster authentic discovery and learning in our students. From elementary classrooms to high school labs, from museum field trips to community-based projects, sensemaking offers a powerful framework for engaging students in authentic scientific inquiry and engineering practices.

The journey from traditional, content-driven instruction to student-centered sensemaking isn't always easy. It requires us to rethink our roles as educators, to embrace uncertainty, and to view our classrooms as dynamic spaces of discovery rather than as venues for information transfer. But the rewards are immense. When we prioritize sensemaking, we're not just teaching science—we're cultivating scientific thinkers, problem-solvers, and innovators. Remember, this shift isn't about discarding everything we've done before. Many of the labs, activities, and resources we've used for years can be repurposed and reimagined through a sensemaking lens. It's about changing the sequence, the framing, and the ultimate goal of these teaching and learning experiences. Instead of using a lab to confirm what students have already been told, we use it as a springboard for questions, investigations, and deeper understanding.

As we move forward, let's challenge ourselves to continually reflect on our practice as science educators. Are we truly putting phenomena and problems at the center of our instruction? Are we leveraging the diversity of our students' experiences and perspectives? Are we creating spaces where all students can see themselves as capable STEM thinkers and doers? Sensemaking isn't just about a better method of STEM education. It's about preparing our students for a world that demands critical thinking, collaboration, and the ability to tackle complex, multifaceted challenges. By embracing this approach, we're not just teaching science content. On a more profound level, we're empowering our students to make sense of their world, to ask meaningful questions, and to use evidence and reasoning to find answers.

In the chapters to come, let's carry forward the excitement and potential of sensemaking. Whether you're a veteran teacher or just starting your career, whether you're in a well-equipped lab or a resource-strapped classroom, there are always opportunities to incorporate sensemaking into your practice. Start small. Be patient with yourself and your students. Remember that like science itself, teaching from a sensemaking orientation is an iterative process. The next chapter focuses on a number of tangible examples you can use to kick off the sensemaking process in your classroom and in other educational settings. The examples are meant to be helpful in terms of supporting the shift to sensemaking in a number of scenarios, such as using sensemaking with your K–12 students; or teaching about sensemaking with college-level preservice teachers; or leading professional development sessions in which you are helping inservice teachers navigate the shift to sensemaking. Personally, I have used the examples presented in Chapter 3 countless times across various groups and scenarios. I personally stand behind their effectiveness in conveying the power of sensemaking. They're also really fun, so get ready to dig in.

CHAPTER 3

Pedagogical Strategies for Facilitating Sensemaking

Facilitating sensemaking requires a different kind of approach from those many of us experienced as learners in the K–12 educational system. These differences are subtle, yet key to helping students become the drivers of their own learning rather than passive containers of information. In this chapter, I aim to make some of these nuances more apparent by providing a number of examples of what I call *mini-reasoning experiences*. In fact, you can use one or several of the upcoming examples right away so as to start implementing sensemaking into your own work, no matter what your role is in education.

These mini-reasoning experiences come from almost 10 years of my own experience working in live television. During that time, I co-hosted a segment of a show called "Science Sunday" on a local CBS affiliate. At the time, I was working at a science center where I would be given a topic from our marketing department, a topic that I needed to turn into a demonstrative science experience. We would film live, in-studio, with the segment to be aired during the Sunday morning broadcast. The topics from the marketing department were usually related to something that the science center was promoting, like an event related to the movie *Frozen* or a Mother's Day special. My job was to think about how to *do science* in connection with these events or topics. In the beginning, this was very difficult and took me a really long time to figure out. Over time, though, I got better at being able to find the science in almost everything. It even became a tongue-in-cheek joke to end these clips with my saying, "Remember, always science . . . ," because we were helping viewers to find the science in everything around them. I also approached filming these mini-reasoning experiences as my way of providing teachers with clips they could use in their own classrooms. So I explained my aim as being the "anti-Bill Nye," a statement that was hilariously taken out of context in one article. I am not anti-Bill Nye. Rather, I meant that I was not trying to steal students' *aha* moments of individual discovery by giving away the answers during my filmed sessions. Instead, my aim was to provide students with just enough information to begin to make sense of it all on their own. I felt and still feel that Bill Nye did a great job of getting many students interested in science. However, I have been critical of the way in which Nye often framed his explanations as equal to doing science—in contrast to spreading the joy of figuring out things on your own. I always knew when a particular airing of the show had been successful. For example, when my wife and I would head to the local grocery store on Sundays after "Science Sunday" aired, people would run up to us and share that they were at the store to buy the same items I had used so that they could re-create the televised experiment in their own homes with their families.

Over time, I came to see the benefit of using these mini-reasoning experiences in the education sector. For example, I could use these experiences in professional learning sessions, in conference workshops, and in my methods courses at Boston University's Wheelock College of Education & Human Development. Rather than simply telling teachers and students about the vision in the Framework and NGSS, I could immerse them in an experience that would demonstrate the kind of learning called for, while also modeling the pedagogical moves that support sensemaking. In her classic work *The Sense of Wonder*, Rachel Carson (1956) wrote, "If a child is to keep alive [their] inborn sense of wonder . . . [they] need the companionship of at least one adult who can share it, rediscovering with [them] the joy, excitement, and mystery of the world we live in" (Carson & Lee, 1956, p. 45). With this quote, Carson effectively captured the idea that adults need to foster and continue developing their own natural sense of wonder so that they can help others find joy, excitement, and mystery in the vast array of phenomena in the world and around them. When I use these mini-reasoning experiences with educators, the energy in the room shifts from business as usual toward joy and excitement—visually apparent as their faces reveal furrowed brows, narrowed eyes, and pursed lips, often accompanied by a tilted head and distant gaze. As these professional educators begin to dig into the topic at hand, their *sensemaking mode* might also include hand gestures like touching the chin or scratching the head. Some might feel an immediate desire to start talking with those around them about what they are noticing and wondering. All of these physical expressions and gestures convey a sense of deep concentration and puzzlement. Thus, by shifting the orientation from "learning about" to "figuring out" science, the educators are able to experience sensemaking firsthand. Ultimately, the experience of immersive learning from a sensemaking orientation feels like the creation of a much-needed space in which we can collectively agree that this is the kind of learning environment to which all students should have access.

Over the years, I have been asked to document these mini-reasoning experiences countless times by teachers with whom I have worked in professional learning settings. I have also been asked by other educators and PD facilitators to document the use of mini-reasoning experiences for their university courses or professional development settings. With the same objective in mind for this chapter, I present the following examples in a simplified manner, with emphasis placed on the following categories: (a) strategies for selecting the topic of investigation, (b) how to introduce the topic, (c) the pedagogical goals for the experience, and (d) tips and tricks for success.

KEY STRATEGIES TO INTRODUCING MINI-REASONING EXPERIENCES

Before diving into these examples, it is important to know a few things about how to successfully facilitate these kinds of experiences with learners of all ages. I have found that the strategies are very simple to understand, yet they take a lot of practice to implement successfully with learners. I think this is because many educators, myself included, were taught in more traditional ways, much like the learners we are working with today, who are extremely experienced in *school as usual*. This means they are great at answering IRE initiation-response-evaluation (IRE)-style questions that only rely on memory recall and privileged facts. Many of the educators with whom I now work are like me in that they were good at the kinds of schooling practice where you are simply told what to think, but many are unfamiliar with sensemaking experiences in which the answer/explanation does not immediately follow the demonstration of a cool science event. In fact, almost every time that I have used sensemaking practices with learners, their first reaction is, "Cool! Why does that happen?" Take that response as an opportunity to engage students rather than an opportunity to explain plain facts. I next share a few common challenges that you can expect to encounter when trying to facilitate mini-reasoning experiences with your students; and in turn, how to react to certain challenges in order to ensure that mini-reasoning experiences will be fruitful in supporting sensemaking for all learners.

Facilitator of Learning

The first challenge is shifting your role from teacher of content to facilitator of learning. This necessary shift comes up early and often when leading sessions covering mini-reasoning experiences. The nature of this challenge makes sense because we, as the facilitators, are usually the ones leading the class or PD session. As such, learners see you as the one who needs to make sense of the concept for them. This is where the "Why does that happen?" question comes up, and how you respond can be really important. I have found that it's important *not* to say something like, "Well, I know the answer, but you need to figure it out," as if there is some kind of magic trick at play and not a real shared experience happening in front of all of us. To enable sensemaking, I often introduce mini-reasoning experiences using generic and everyday language, a strategy that helps to lower the stakes and get everyone involved right away. To help explain this more fully, I next use the ring and chain loop as an example.

The ring and chain loop consists of a small metal ring (1–2″) and a normal metal chain—like a cheap necklace chain—used together to create a perplexing event. When I introduce it, I say something like, "I want you all to check this out because I was playing around with it the other day and noticed something a little strange that *sometimes* happens." Then I let the chain dangle over my fingers (with most of it hanging down) and bring the ring up the length of the chain so it is resting near the top, but on the outside of the chain. I ask learners what they think will happen if I just drop the ring. Depending on how well my students know me—that is, if I have done other things like this with them before—students will either say that the ring will just fall to the ground or it will get stuck in the chain. I think these responses are worth noting because they show that learners are trying to predict something anomalous to what they expect, rather than what they truly believe will happen. No matter what the prediction is, I ask students to share their reasoning behind *why* they think their particular prediction will occur. Once we have all the options on the table, I let the ring fall and see what happens. If the group, as a whole, tends to agree that the ring will get tied up in the chain, then I often let it fall to the ground a few times and acknowledge that this is all that seems to happen.

Poker Face

Before continuing with the ring and chain example, I want to highlight a second strategy that is extremely important in facilitating these kinds of sensemaking experiences: the poker face. The poker face is key when it comes to leading sensemaking experiences because learners of all ages are well accustomed to reading body language, especially when they want to see if they are on the right track with their responses when trying to figure out reasoning experiences. I have worked with so many teachers and facilitators who have not initially realized that their expressions and gestures *give it away* (the answer) when leading sensemaking experiences. Some tend to nod their heads "yes" when learners are on the right track. Other facilitators often make confused facial expressions when incorrect responses are given. As for my past work as a facilitator, I would smile and nod when students were on the right track. However, after I began working on keeping a poker face, I had a moment with a group of 3rd-grade students who asked me a question, and I was able to keep a straight face. Then a little girl said to her elbow partner (classmate nearby), "My question was so good, even Mr. TJ doesn't know the answer!" I remember the moment fondly because I felt like I had finally made some progress on not wearing my thoughts on my face!

Now, back to the ring and chain example. While dropping the ring around the chain and to the floor a few times, learners will undoubtedly be looking at your face, thinking, *This is what I expected to happen, but he wouldn't be showing us this if something else didn't happen!* Keeping a poker face is important in this instance, and I usually say something like, "That makes sense, right? It should just fall, right? Well, sometimes I notice a weird thing happens." The next time I drop the ring, I shift the ring a little to the side so that this time when it falls, the side of the ring slides down the chain with more contact as it falls. If done correctly, the chain wraps around the ring, tying it in a loop. The ring does not fall to the ground; instead, it hangs as if it were a part of the necklace, dangling from the chain. Again, imagine yourself doing this as a facilitator. You would notice that students immediately look at your face to see if you make a *gotcha* face, thereby revealing that this was your intention all along. Once again, it is key to keep the poker face. Moreover, I have found it to be incredibly useful to position yourself as a part of the sensemaking community by making a confused face, then pausing to acknowledge the confusion at what just occurred with the question, "Wait, what just happened that time?" In response, if one of the learners states aloud that the ring should have fallen to the ground—instead of somehow tying a knot in the chain—the statement would serve as helpful feedback that everyone in the group could hear, also serving as evidence of someone in the group going public with what they noticed. After demonstrating that the ring sometimes does get tied in the chain, you might expect several learners to typically reach for the ring and chain, as they want to try it out on their own. This is a helpful occurrence because you can hand the ring and chain off, and while someone else tries it, you can position yourself as part of the group, moving around the room and talking with everyone as they begin to reason about what is happening.

Ring and Chain

Strategies for Acquiring the Item:

- Searching online for the "magic ring and chain trick" will be the easiest way to find these two items. They are usually only $1-$3 each. Every once in a while, you can buy them in bulk cheaply, which can be helpful if you are using this strategy with a large group (i.e., for professional development).

How I Introduce It:

- Included in the surrounding text.

The Pedagogical Goals:

- The main goal with the ring and chain activity is to focus in on the following:
  - » CCC—Cause & Effect
  - » SEPs—Asking Questions, Constructing Explanations
  - » DCI—Forces and Interactions
- This mini-reasoning experience is a *low-floor, high-ceiling* type task that makes everyone curious and engaged. It does not take any time to set up. The aim is to have students immediately asking questions, making observations, and starting to construct initial explanations about why the phenomenon is happening. It has a high ceiling because it is complex to mechanistically explain and seems so simple when you observe it for the first (or 100th) time. I have even used this strategy when explaining phenomena to a PD group in California, including PhDs representing their content areas. In addition, a high school physics teacher had so much fun trying to explain the phenomenon that he noted, "This brought me back to what it feels like being a learner again, and makes me rethink how I should introduce a lot of my topics to my students!"
- Depending on how much time you want to spend on this lesson or training, you can include the SEP Planning and Carrying Out Investigations, Modeling, and/or the CCC Structure and Function to move into trying to create a mechanistic model of what is actually happening. Students often want to record the phenomenon with their cell phones on slo-mo. Doing so is really interesting because the video is usually not sufficiently compelling evidence to support what actually happened. As a result, there is still room for the learners to try to explain what was making the ring get stuck on the chain.

Tips and Tricks for Your Success:

- When using the ring and chain activity with a class of K-12 students, I typically use one set so that the activity does not turn into something that everyone wants to try on their own. Instead, the one-time event keeps our collective focus on trying to figure out what is going on.

- When students are using the ring and chain, you can shift your role to asking them, "How did you do that?" This approach is often very productive because the movement with your hand is so subtle that they usually don't know the answer. This can be helpful in starting to put together an explanation for how it happens.
- Don't be surprised if the ring falls more than expected, or if it gets stuck more than expected. I have found it very helpful to maintain my own poker face when I do not know what is going to happen each time. We all can join in on the wonder!

One final and key detail about the use of a poker face is its potential to shift the agency in the classroom over to students—that is, positioning the learners to do the heavy lifting of sensemaking. Everyone, at some point during their years of schooling, has had the classroom experience in which they responded with an answer to the teacher's question. They either saw the teacher's questionable facial expression or received a "Not quite" response, immediately knowing they were on the wrong path. In a classroom where sensemaking is the focus, the teacher assumes a key role in the learning community. Maintaining a poker face when students are going public with their reasoning seems small, yet it is a huge part of shifting the power dynamics of our classrooms to students. Over time, students become empowered by this shift to take on more responsibility in their use of the science and engineering practices and the crosscutting concepts. In fact, many students even refrain from looking up the right answer or will call out others to prevent them from *ruining it* (i.e., the fun of figuring things out) in order to explore topics in an authentic way, without someone stealing their aha moments!

IMPORTANCE OF MINI-REASONING EXPERIENCES

To some, mini-reasoning experiences often seem like fun activities for students to engage in as a learning bonus, or as an "If we have time" thing that would be nice to add beyond the regular lesson. I do not see them this way at all. I have used mini-reasoning experiences with students and teachers countless times and have come to realize their importance over the years. One of the most powerful aspects of using these experiences is their ability to enlighten students on what the STEM field is all about—that is, using sensemaking to investigate phenomena and solve problems. In contrast, almost everyone with whom I have worked over the years thinks of science as being more about remembering facts that were figured out and made public by other people, as opposed to understanding science as active engagement in one's own sensemaking. I see this range of attitudes toward science education every semester when I do an *AutoSTEMgraphy* assignment in which each student graphs their own lifetime of STEM experiences, with time on the X-axis, then plotting each STEM experience on the Y-axis, with the top being very positive and the bottom being very negative. For example, a student might put a field trip to an aquarium as a positive experience in middle school. Yet that same student might list a high school chemistry course, in which they just had to memorize the periodic table, as a negative experience. Of course, most of us could imagine the kinds of STEM experiences that students would consider positive (e.g., field trips, science fairs, fun experiments) versus other STEM experiences that they would likely consider negative (e.g., textbook-heavy assignments, worksheets during COVID-19, rote memorization). Interestingly, one of the emerging trends noted from this activity indicates that students rate challenging STEM courses very positively if their teacher provided them with the support and encouragement needed to challenge themselves and experience success. You might think that hard courses would naturally be rated as *less positive* when compared to a field trip or fun project that students could choose to pursue. When I have students share their STEM experiences with classmates, they often describe the more difficult STEM courses as being so positive because their teacher saw something in them that they did not necessarily see in themselves initially. Overcoming the challenges and seeing your own accomplishments is a rewarding experience for students, one that builds confidence in tackling new and challenging experiences in the future.

When you are introduced to a mini-reasoning experience, it often feels like a trick. We can see that something is happening, but we are not sure where to begin with making sense of what actually is happening. This issue became a running joke for my "Science Sunday" television segments, whereby people would see me out in the world and say, "Hey, science guy, do some of your tricks!" This was so much the case that we would make sure to talk about the fact that there are no tricks involved. In fact, it's just science that we haven't figured out yet. Nonetheless, one thing that these reasoning experiences and magic tricks *do* have in common is that people like them and are engaged enough that they want to stick around and figure out what is happening. In one of my first-year courses at Boston University—STEM Learning, Action, and Impact—I promise students that we will do science every day (i.e., a mini-reasoning experience). I have found that it only takes one class session for the students to enthusiastically call for more. For students, the power of mini-reasoning experiences centers on being high interest and fun to engage. There is also an inherent challenge to these sensemaking experiences that feels rewarding to figure out. Thus, the more experience students get in sensemaking, the more competent they feel when trying to figure out the next one. While I usually only refer to these mini-reasoning experiences as "time to do science" or "time to play around with science," I cannot help but reflect on my belief that the shifts I see regarding students' approaches to sensemaking mirror the simplest definition of learning: lasting behavior change through repeated practice. Students shift their behavior away from not liking STEM to feeling confident and successful at sensemaking, one experience at a time.

HIGHLIGHTING THE THREE DIMENSIONS WITH SENSEMAKING

The gap between the three dimensions of the NGSS and where to start with students or teachers can seem a mile wide. One potential starting point is to start with a mini-reasoning experience and slowly unpack each dimension in collaboration with your learners, in this way helping them experience what it *feels like* to engage in three-dimensional science teaching and learning. To provide another example of how to do this, I next introduce the beaded cup activity as a way to experience three-dimensional learning that works well every time. Moreover, learners love the activity and will ask you to do it over and over and over.

The beaded cup is a clear plastic cup (with a handle) that contains a single chain of beads (picture Mardi Gras beads) that is 50 feet long. To get the setup started, I pull the chain of beads from its box and get the cup before students arrive to class, making sure to set up the activity out of sight. Setting up these components requires dumping the long chain of beads onto a table, finding one of the ends, and then feeding the chain carefully into the cup so that it does not get tangled along the way. The first couple of times you do this, it might seem like it will take forever because you feel like you need to be more careful than is actually necessary. On the other hand, if you stop paying attention to how you are feeding the beads into the cup, you will likely find that as the cup gets increasingly full of beads, the beads will automatically start spilling out of the cup before you are ready. When you get to the other end of the chain, the cup will be full to the top. You will want to carefully position the end of the beads in a spot where you can grab them.

When introducing the beaded cup activity to students, I usually say, "I found this thing and want you all to check it out. Let's start with what it is. What do you notice about this?" Students share that it is a clear cup that looks like plastic, maybe, with beads in the cup. Then they begin to ask about the beads. When they ask if the beads are all connected, I share that they are one long chain of beads. I ask them to think about what would happen if I hung a few of the beads over the edge of the cup and took my finger off? At this point, I am holding the cup by the handle and walking around so everyone can see a couple of the beads hanging over. The group is usually split, with some thinking that something *must* happen (or why would you be showing us?), while others think that the rest of the beads are in the cup, so nothing should happen. No matter what perspective is shared first (remember the poker face), I always respond with, "Okay, what else could happen?" This ensures that multiple students get to share, and multiple possibilities are made public. Sometimes the entire group is in agreement that all the beads will come pouring out of the cup as soon as I lift my finger off the few hanging over the side. If this should happen, I pose a question asking students to consider how this could happen. I say, "The vast majority of the beads are inside the cup. I'm not sure what the number is, but if there are only three to five beads hanging over the edge, then there have to be hundreds and hundreds still inside the cup. So how would those all start falling up and out over the edge of the cup?" This combination statement/question usually sways a few students, but if others are still unsure, I sometimes give an analogy to get people thinking—usually starting with a disclaimer that this is not a perfect analogy. What if we thought of the cup as an aboveground swimming pool and we are all the beads, holding hands in one long chain? If a few of us went out (i.e., over the edge of the pool), would the rest of us get yanked out over the edge, too? Usually, this analogy results in students' facial expressions indicating that they are now realizing that the majority of the weight is in the cup. Therefore, it does not make sense that the beads would pour out. At the same time, some of the other students' expressions indicate that they don't buy these scenarios as being analogous. Either way, it is important to get everyone to understand that the majority of the chain of beads is in the cup, and that it would make sense for those beads to stay in there—even if a few were dangling over the edge.

Once all positions have been stated and reasoning has been shared for why the beads will either fall out or stay in the cup, I tell students that I am going to lift my finger off the beads. It is important to note that at this point, you are holding the cup by the handle and extending your arm straight out at shoulder height as you lift your finger off of the beads. Sometimes students want to know exactly how many beads are over the edge, so you can count them together as you put each over the edge and lift your finger each time. Eventually (around 6 beads or so), the beads start to rapidly pour out of the cup and onto the floor. Surprisingly, the beads do not just drag along the edge of the cup, but rather seem to go straight up into the air, turn around, and then fall straight down to the ground. This happens until all the beads are out of the cup and on the floor. At this point, I ask students to share what just happened. They tend to share a number of observations. Sometimes it takes a few people initially sharing their ideas before other students make note of the fact that the beads didn't just pour out, but actually went up and then down. Usually this perspective is shared by a student with an unsure face, almost as if they saw it but do not believe what they saw with their own eyes! Whenever students are sharing, I make sure to respond with "talk moves" to ensure that I heard what they were saying clearly, and also to ensure that everyone else in the room understood their contributions as well.

The one thing that I can guarantee is that everyone will want you to do the beaded cup activity again. So I gather the beads and begin loading them back into the cup while encouraging students to share their thinking about what had occurred the first time. While their ideas are usually all over the place, I try to position myself as a member of the group who is also trying to make sense of this phenomenon. As such, I agree with any ideas shared and make sure that my comments are shaped in the form of questions. Alternately, I share my agreement that this scenario is confusing, that it does not make sense to me just yet, but that any contributions from students are helping to make sense of it all. At this point, I have the cup and beads reset and tell students that I am going to repeat the activity. I even tell them to *feel free to record it* because at least a few students will want to have their phones out to record the event on video or by using their slo-mo capture. While I do not require any students to film it, letting students know that recording is permitted usually saves you a third reset. At the same time, this process is setting you up to have students suggest a slightly new experiment, using the same cup and beads, but in a different way.

Beaded Cup

Strategies for Acquiring the Item:

- I have purchased this item a number of times from an online company called Educational Innovations. The item is listed as a Polymer Bead Demonstration.

At the time of this writing, it is listed in the chemistry section under molecular models. There may be other companies who sell this product, but the key is to search for *Polymer Bead Demonstration*. Once you find it, you should see a clear plastic cup with a single chain of beads (often red, white, and blue).

How I Introduce It:

- Included in the surrounding text.

The Pedagogical Goals:

- The main goal of using the beaded cup activity is to focus on the following:
  - » CCC—Cause & Effect, Scale, Proportion, & Quantity
  - » SEPs—Asking Questions, Planning and Carrying Out Investigations, Constructing Explanations
  - » DCI—Forces and Interactions
- This mini-reasoning experience is great because it is so simple and repeatable. It immediately prompts students' thinking and then asking, "What would happen if . . . ?" This response demonstrates our innate desire to plan and conduct investigations, a key science practice in the NGSS. Students wonder what would happen if the cup were on the floor or positioned up high. They are able to immediately see evidence of changes that support their own sensemaking about why this phenomenon occurs.

Tips and Tricks for Your Success:

- After holding the cup at arm's length, students will think that the beads will always fly up and out of the cup. But when prompting them to see what happens when you do it on a table or on the floor, they become even more perplexed upon seeing that the beads do not fly up anymore. They even stop after a short while. This usually causes students to wonder if gravity is having an effect, and they ask to try it while holding the cup as high as possible. With the student (safely) standing on a chair or other raised surface, the effect seems to happen even more drastically, with the beads flying even higher up and out of the cup before falling back to the floor.
- As students observe these differences, the topic of gravity being stronger or weaker at different heights almost always comes up. Then, it is often helpful to ask if gravity changes between the floor and the ceiling of the room. "Yeah, I totally see that, but wait. Would gravity really be *that* different from the floor to the ceiling here in the room we are in? How could we determine if it was just the gravity changing, or if it was actually something else causing this to happen?"

When I use the beaded cup, my aim is to show learners something perplexing that ought to be quite simple given the materials at hand. Experiencing this activity encourages students' minds to go right to planning and carrying out investigations because of the natural desire to see what would happen in slightly different scenarios. As students explore these variations, the concept of gravity's strength differing with altitude often emerges. Students may ask (or you might seed the idea), "What would happen if we did this on the floor?" When the beads start coming out on the floor, they sort of roll over the edge (not flying up out of the cup) and stop after a short while. This usually stumps students. Why would it stop? What is causing it to stop? Ideas about gravity typically emerge, and students usually ask, "What happens if you go even higher?" Standing on something tall (safely) dramatically changes the situation, and the beads seem to jump up even higher out of the cup. Students also share that the beads seem to go higher at this position, seemingly impacted by gravity. You might follow up and ask, "Interesting observation, but consider this: Would gravity truly vary noticeably between the floor and ceiling of our classroom? How could we distinguish between actual changes in gravity and other factors affecting our experiment?" This mini-reasoning experiment shows how easy it is to begin to make sense of these phenomena by changing simple variables while maintaining everything else that is going on.

Another important pedagogical move tends to come up when you use phenomena like this one, and it is spurred by the following question from learners: "Okay, so are you going to tell us the answer?" How you respond to this question is key and ultimately depends on your goals and the type of student group with which you are working. I will describe three different response options, along with my thoughts about these responses in relation to the learning goals set for each group.

1. If I am working with a group that is entirely new to this kind of learning, I usually tell them, "I don't want to steal your aha moment." Then I leave the explanation wide open for them to think about on their own. While learners will usually groan at this response, I think it is crucial to help them become comfortable with being uncomfortable, as not knowing the explanation of something that seems so simple can be a tricky feeling and takes some getting used to.
2. If I am working with a group that is new to this type of learning, but I know that I will see them again (e.g., methods students, multiday PD participants, or K–12 learners), I also tell them, "I don't want to steal your aha moment," and I still prompt them

to pursue ways of making sense of the phenomenon on their own. I encourage them to talk to others—such as friends, family, and community members—about what they saw and to practice facilitating this conversation by using talk moves to support the discussion. If the phenomenon is something they can show by way of a video recording or the item itself, then I tell them to try out their own poker faces while leading the sensemaking discussion to see where it takes them. Depending on the group, I might also encourage them to resist the urge to Google to find more information or answers. In certain situations, I might leave the Google option open so that we can have a conversation about what it feels like to just know the answer. This particular conversation often involves students sharing that it feels unrewarding to get answers via Google; also, their Google searches often leave them with more questions than they started with before the search. One quick note: Many scientific explanations that students find online are full of science vocabulary, and they do not get at the conceptual understanding or mechanistic explanations for these phenomena.

3. In the rare case when I do share the answer or a partial explanation with learners, I usually do so to illustrate the idea that it is much more rewarding to figure things out on your own rather than have someone tell you why the phenomenon occurs. It is usually compelling to draw an analogy likening being given the answer in a scientific exploration to going to see an incredible magic show during which the magician tells you how the trick happens every time they do it. The feeling of wonder and curiosity, along with the journey of sensemaking, is much more rewarding than immediately knowing the answer. The more students engage with phenomena, the more they will come to understand and appreciate the sensemaking process.

Sensemaking experiences like the beaded cup demonstration offer a powerful entry point into three-dimensional learning as outlined in the Next Generation Science Standards. By engaging students in a perplexing phenomenon, we naturally draw them into the practices of science and engineering. Students begin asking questions, planning investigations, and constructing explanations—all key elements of the Science and Engineering Practices dimension. As they grapple with the unexpected behavior of the beads, they encounter and apply crosscutting concepts such as cause and effect, scale, proportion, and quantity. These concepts help students bridge their observations to foster deeper scientific understanding. Finally, as students explore variations of the demonstration and discuss possible explanations, they begin to uncover disciplinary core ideas related to forces and interactions. This approach to learning science mirrors the authentic work of scientists, positioning students to experience how scientific knowledge is constructed rather than simply consuming facts. By anchoring instruction in phenomena that spark curiosity and demand explanation, we create an environment where all three dimensions of the NGSS can naturally intertwine, leading to more meaningful and lasting understandings of scientific concepts and practices.

MOVING FROM EXPLAINER TO FACILITATOR OF LEARNING

A major challenge of the shift to sensemaking in K–12 classrooms is the transition of the role of the teacher from *explainer* to *facilitator of learning*. This topic has come up multiple times in this book, as it is a complex topic viewed from a number of perspectives. The first and most important challenge is the shift in the relationship between teacher and learner. Learners expect the teacher to teach them things by feeding them prescribed information. I have experienced this dynamic from the prekindergarten level all the way up to the graduate level, and it is not a simple shift. We do need to teach students things. However, as educators in the STEM field, we now know more about how people learn. Thus, we have updated the ways in which we support students in their own learning processes. In particular, we have made the shift toward a deeper understanding of the fact that simply telling students information does not lead to real learning. At the same time, we are not expecting students to come up with everything on their own. The Ambitious Science Teaching team from the University of Washington, led by Mark Windschitl, has been one of the most helpful resources for addressing the teacher's role as a facilitator of learning. I first encountered this teaching approach in a 2015 primer titled *Just-in-Time Instruction*, written by the Ambitious Science Teaching team. Later, Windschitl et al. (2020) expanded on the primer's content with their book *Ambitious Science Teaching*, in which the authors referred to the concept of just-in-time instruction as "Interactive Direct Instruction" in the chapter titled "Introducing New Ideas" (p. 151). Next, I provide a short description of this kind of instruction followed by an example of the process in practice.

Just-in-time instruction (aka Interactive Direct Instruction) serves as a powerful tool in supporting student sensemaking in science classrooms. By strategically introducing conceptual ideas at moments when students feel a genuine need to know more, this approach bridges the gap between students' current understanding and the knowledge required to progress in their scientific reasoning. Unlike traditional lecturing, just-in-time instruction responds directly to students' emerging questions and puzzles, providing students with carefully chosen conceptual frameworks that they can immediately apply to the phenomena they are investigating. This method of instruction aligns closely with how scientists actually work, encouraging students to grapple with ideas and use new information to construct explanations. When a teacher introduces a concept like buoyant forces—just as students are about to explore why objects float or sink—it gives these learners the conceptual language and ideas they need to begin making sense of their observations. The key is to provide just enough information to enable students to reason and construct explanations on their own, rather than delivering a complete explanation that would rob them of the opportunity to engage in authentic scientific thinking.

By timing the introduction of new ideas to coincide with students' investigations, just-in-time instruction makes abstract concepts more concrete and immediately applicable. This approach helps students see the relevance of the new information, increasing their motivation to understand and use these ideas in their own reasoning. As students apply these newly introduced concepts to explain phenomena or solve problems, they engage in deeper levels of sensemaking by connecting the abstract with the concrete and developing a more robust understanding of scientific principles. Additionally, just-in-time instruction supports sensemaking by acknowledging and building upon students' prior knowledge and experiences. By framing new concepts in relation to what students already know or have recently observed, teachers can help learners integrate new information into their existing mental models. This process of connecting new ideas to prior knowledge is fundamental to effective sensemaking and helps students construct more coherent and scientifically accurate understandings of the world around them.

Using an example might be the best way to illustrate this kind of instructional approach. In more traditional lessons, you would typically first teach about the properties of air and kinetic molecular theory before giving students a lab experience to confirm their understanding of what was just taught. A sensemaking approach would be to give students a phenomenon or challenge to investigate. Then you would allow them to experiment with it in order to drive questions that would lead them to *need* to know more about the properties of air in order to explain this phenomenon. One approach to do this is through the use of an item called a windbag. Windbags are 8-foot-long, very narrow plastic bags that arrive neatly rolled up in a way that makes them seem much shorter than they actually are. As an activity that works well with groups, you would give each group of students their own windbag and challenge them to find *the fastest way to inflate their bag* with the help of all the group members. Keeping this challenge seemingly vague is important because each group needs to determine their own criteria and constraints in terms of how they will decide what is the "fastest way" to fill their windbag. As a facilitator, your goal is to create the conditions that lead to the best learning outcomes. In this case, you are leading students to think about air in new ways. This is a challenge because air is something so commonly encountered throughout our lives that it can be tough to imagine a new way to think about the topic of air.

The activity begins with the groups of students unrolling the windbags, noticing that both ends are open (i.e., not tied or sealed). On this point, students usually come right up to me to ask why this is the case. Shrugging my shoulders, I simply repeat the challenge: "Find the *fastest way to inflate the bag* with the help of your group." At this point, some groups are running around the classroom trying to fill the bag like a windsock. Others tie a knot in one end and then start taking deep breaths, blowing into the other end like a balloon. This is another scenario during which you can practice your poker face and your facilitation skills—moving around from group to group and asking students to share the reasoning behind their decisions. As groups begin to fill the bags, which will take a while using these two approaches, I shift to asking students, "Okay, but is this *the fastest way to inflate the bag*?" The question prompts the students to consider alternative approaches to completing the challenge. It usually takes about 20 minutes or so for each group to have enough time to experiment with multiple approaches before being ready to share their ideas. At this point, I typically move around the room, allowing each group to present their approach. The class always seems to gain confidence as each group shows similar ways of thinking.

At this point, I take a windbag of my own and tell the group about a way to inflate the windbag that uses a special property of air that we should explore further. I tie one end of the bag and invite a student to come to the front of the room to hold the tied end. Holding the bag parallel to the ground, level with the height of my mouth, I open the end of the bag that is facing my face and hold the opening around 6 inches from my mouth. After taking a deep breath and blowing a burst of air into the bag, a ripple of air can be seen moving through the length of the bag before it seems to fill from the opposite (tied) end of the bag. I grab the open end to seal it once the bag has filled with air. Students usually want to rush to try this strategy with their group's windbag. Before you let them, ask the students to think about *why* the bag fills so quickly this way, as compared to the other approaches they tried. Once groups begin having success with this new approach, I bring around chart paper and instruct each group to create a model (with pictures and words) that demonstrates why this way of filling the windbag is faster than the other approaches they used to fill it. It is important to instruct them to represent both the parts of the windbags that they can see and the parts that they cannot see (i.e., the air) in their explanations.

Having students engage in modeling at this point is important because it creates the need to carefully explain what they can see. It also encourages them to conjecture about the unseen variables at play in this phenomenon. Moreover, when you are in a modeling experience like this, you reach a point at which it is much easier to ask questions about what you are unsure of than to present components of an explanation about which you are also unsure. So, once I begin to see the confused expressions on my students' faces, I pause the class to provide additional instructions and resources. Specifically, I let them know that while they are creating their models, they can use sticky notes (e.g., Post-it notes) to write down their questions—one question per sticky note—and place the notes at the bottom of their chart paper. While you will likely notice that the questions come easier to students when using the sticky notes, I recommend that you still encourage students to draw what they think is happening in the two scenarios (i.e., one breath vs. slower methods). Next, it can be helpful to remind students that at this point, it is not their job to be correct. Instead, they should focus on using the evidence they can see, combined with their own prior knowledge, to try to explain what is happening with the windbags.

Once students have had a chance to share their models, focusing on explaining why one way is faster than the other, make the shift to having students share their questions. As they read questions off the sticky notes, students can post the questions—one by one—on a board for the whole class to see. If a question matches or is very similar to another that has already been shared, encourage students to group those questions together on the board. In this way, the driving question board allows students to see the natural categories forming around their questions, such as the movement of air, air pressure, forces, the shape of the bag, and the role of the quick breath versus the air in the room. This is an important step because students will be able to recognize that their questions are similar to ones asked by other classmates. They will also need to find a way to explain these similarities. Furthermore, creating and maintaining a DQB—a visual representation of the sensemaking process that everyone can see while carefully listening to the questions being shared—takes a substantial amount of time because students are also collectively thinking about how to identify the categories for their questions. These are all important conversations and should not be rushed, as spending time here on the front end tends to speed up the sensemaking process later on.

Windbags

Strategies for Acquiring the Item:

- I have purchased One Breath Bernoulli Windbags a number of times from an online company, SteveSpanglerScience.com, run by Steve Spangler. The bags are bright, colorful, and designed specifically for use in science instruction.
- While I have used other bags for this activity in the past (e.g., diaper bags), I have found that they do not work as well as the windbags. In addition, such substitute bags are often clear and much less fun.

How I Introduce It:

- Included in the surrounding text.

The Pedagogical Goals:

- The main goal of working with the windbags is to focus in on the following:
  - » CCC—Cause & Effect, Scale, Proportion, & Quantity
  - » SEPs—Asking Questions, Planning and Carrying Out Investigations, Constructing Explanations, Developing and Using Models
  - » DCI—Matter and Its Interactions, Motion and Stability, Energy
- This mini-reasoning experience is great for a group challenge and does well in bringing teams together in lieu of a random team-building game. I like to use the windbag activity at the beginning of the year in order to help set the stage for the kind of learning that will occur as we move forward in the course. Another goal is to help students see that STEM is enjoyable when it is not just about remembering facts. Overall, I want to help my students experience STEM education as an opportunity to use their current knowledge to explore and engage increasingly more complex ideas.

Tips and Tricks for Your Success:

- Practice inflating the windbag a few times before doing the activity as a group demonstration. If you have no one around to help hold the other end of the bag while you practice, you can use a classroom cabinet or locker to hold that end of the bag at mouth height.

- If you plan on using these windbags over and over again, tell students not to tie the ends of the bags too tight. Or you can use a rubber band or clip to seal the bag rather than tying it. If students are rough with the bags, which often happens because the bags are very light, you may need to replace them after a few uses to get the best effect.

As you shift to just-in-time instruction for ongoing lessons and units, you might begin by acknowledging students' reactions of amazement at the windbag activity during that particular class session. This kind of reminder also helps students to reflect on what was done previously, and how they engaged with the sensemaking process. You could say something like,

> *I saw a lot of surprised faces when that bag inflated so quickly! Many of you asked how it's possible to fill such a big bag with just one breath. Well, there's a really interesting scientific principle at work here, and understanding it will help us explain all sorts of cool things—from how planes fly to why those windbags inflated so easily. Let's talk about something called Bernoulli's Principle.*

Before you dive into the new stuff, remind students about air pressure. Ask them to recall how air molecules are constantly bumping into things, creating pressure. This helps set the stage for what's coming next. Here is where you introduce the main idea by stating, "Bernoulli's Principle tells us that when air (or any fluid) speeds up, its pressure goes down. It's like the air pressure and speed are on a seesaw—when one goes up, the other goes down." To help them visualize this principle, you might want to draw a simple diagram on the board. Show air flowing through a tube that gets narrower, with arrows indicating that the air speeds up in the narrow part. Add some "P" symbols to show that pressure decreases where the air is moving faster. Now, here's a fun way to help them relate to this concept. Introduce the following scenario:

> *Imagine you're on a busy sidewalk. When you get to a narrow part of the sidewalk where there's construction work going on, everyone has to walk faster to get through, right? Air behaves similarly when it's forced through a narrow space. It speeds up.*

To really drive the point home, you can talk about how this principle makes airplanes fly. Explain that the shape of an airplane's wing makes air move faster over the top than the bottom, creating a pressure difference that lifts the plane. Now, circle back to the windbag activity and explain,

> *In our windbag experiment, when you blew into the bag, you created fast-moving air at the opening. This fast air had lower pressure than the still air in the rest of the bag. And if air can only push, then the air inside the bag isn't drawing other air in. Rather, it's the atmospheric pressure outside the bag that's doing the pushing because the air moving into the bag is creating a space that air naturally moves into.*

It's all about the pressure difference. Get the kids involved by having them turn to a neighbor and brainstorm other situations where Bernoulli's Principle might be at work. Encourage them to use the term "pressure difference" in their explanations. Wrap things up by summarizing the key ideas:

> *Bernoulli's Principle helps us understand how air speed and pressure are related, and the important concepts to remember are pressure and velocity, and the fact that they have an inverse relationship. When one goes up, the other goes down. Don't worry if it's not all crystal-clear yet. We'll be exploring these concepts and principles more in the coming days.*

It is important for science educators to remember that the goal of the mini-reasoning experience is not to give students all the answers. Instead, the goal is to provide just enough information to get them thinking and ready for the next activity. This approach gives students the tools they need to start figuring things out on their own. It also ensures that they have the information necessary to make progress in developing their own scientific understandings as their STEM studies build in complexity over time.

SENSEMAKING WITH TECHNOLOGY

Before diving into the exciting world of sensemaking with technology, let's take a moment to consider a framework that can guide our thinking—the Triple E Framework, which was developed by Liz Kolb, Clinical Associate Professor of Education and Learning Technologies at the University of Michigan's School of Education (Kolb, 2011). In my Educational Technology course, we focus on this framework because it provides a practical approach to integrating technology in education, that is, an approach that prioritizes real learning over the use of flashy new tech items that educational tech companies constantly try to sell to you as the solution to solving problems. The Triple E Framework focuses on three key components: (a) Engagement, (b) Enhancement, and (c) Extension of learning goals. The *Engagement* aspect involves considering how technology tools might motivate learners to focus on the learning goals. When we focus on the

Enhancement component, we look at how technology can support students in developing a more sophisticated understanding of the content. Finally, using the *Extension* component, we examine how the technology can create opportunities for learning outside of the traditional school day or setting. As we explore various accessible technologies for sensemaking, keep these three Es in mind. They'll help us ensure that our use of technology isn't just flashy or fun, but more importantly, that key aspects of technology use can be genuinely supportive of deep and lasting learning in STEM.

When I think about sensemaking with technology in K–12 classrooms, I'm not necessarily referring to high-tech labs or expensive equipment. Some of the most powerful sensemaking tools I use are already in your pocket or your classroom. While cell phone use is becoming more complex as schools create policies to help curb inappropriate use in the educational setting, let's explore how these everyday, accessible technologies (cell phones) could be used to transform the way students engage with and understand complex scientific phenomena. Even if not every student has a smartphone with a slow-motion camera built in, these items are common enough that they can be borrowed or donated for classroom use. The simple slow-motion feature can be a game-changer when you are working to explain phenomena and solve problems. Imagine students studying the concept of a falling Slinky. They hold the top of the Slinky (allowing it to hang down along its full length) with the bottom a short distance from the ground. In the past, we might have dropped the Slinky and observed its movement with our eyes, or timed its descent with a stopwatch—both strategies leading to inconsistent results and frustration. But with slow-motion video, students can capture the fall in exquisite detail, frame by frame. They can actually see that the bottom of the Slinky does not begin to fall as soon as the top is dropped. Actually, the bottom seems to *wait* until the top catches up to it, and then the whole Slinky falls together. Being able to observe this using a cell phone isn't just a cool effect; it's a way of turning a once-distracting item into a powerful sensemaking tool.

Digging into the App Store (Apple or Android) opens up a world of options for using relevant applications to enhance the implementation of sensemaking experiences in the science classroom. Such apps can get really interesting! I have found countless free apps (e.g., PhyPhox, Vernier Video Analysis, VidAnalysis, etc.) that enable students to measure the distance fallen in each slow-motion frame, plot the data, and discover something fascinating about the nature of falling objects. Students might notice that in the first second, an object falls a certain distance—let's say 5 meters. But after 2 seconds, it hasn't just fallen 10 meters. It's actually fallen 20 meters! And after 3 seconds? It's plummeted 45 meters! What's going on here?

What students are uncovering is the quadratic nature of the distance–time relationship in free fall. It's a fancy way of saying that the distance an object falls isn't directly proportional to the time it's been falling. Instead, it follows a dramatic pattern: the distance is multiplied by the square of the time. So when you double the time, the distance becomes four times greater. When you triple the time, the distance becomes nine times greater. If that sounds a bit abstract, that's exactly why the slow-motion video is so powerful. First, students can see this relationship unfold before their eyes, frame by frame. When they plot their measurements on a graph, they won't see a straight line. Instead, they'll discover a curved line called a parabola—the telltale sign of a quadratic relationship. This hands-on discovery brings to life what's usually just an equation in a textbook: distance = 1/2 * g * t^2, where g is the acceleration due to gravity and t is the time. Now, here's the kicker: Students don't need to be told this equation up front. Through their own observations and measurements, they can build an understanding of how falling objects behave in the real world. The slow-motion video transforms an abstract equation into a visible, analyzable phenomenon. And who knows? This experience might spark all sorts of questions: Why does it fall this way? How would it be different on the moon? What if we added air resistance to the mix?

The sensemaking potential of smartphones doesn't stop here. Many phones now come equipped with a variety of sensors—accelerometers, gyroscopes, magnetometers—that can turn the device into a pocket-sized science lab. With the right apps, students can measure acceleration during a car or bus ride, analyze the frequency of sound waves from their favorite artists, or even measure light intensity around their communities. These tools enable students to collect real-world data easily and often. Instead of relying on textbook examples or one-off lab experiments, students can investigate scientific principles in their everyday lives. This kind of continuous engagement with phenomena and trying to understand the world around them through data collection and analysis is at the heart of sensemaking.

Additionally, the world of augmented reality (AR) is becoming increasingly accessible through smartphones and tablets. AR apps can overlay digital information onto the real world by creating interactive, 3D learning experiences. Take the study of atomic structure, for instance. It's one thing to see a flat diagram of an atom in a textbook. It's quite another to use an AR app to project a 3D model of an atom into the classroom, walk around it, and watch as electrons orbit the nucleus. Students can manipulate the model, add or remove protons and neutrons, and see how this affects the atom's properties. This kind of visualization can be crucial for sensemaking, especially when dealing with concepts that are too small or abstract to observe directly.

If you are the type of individual who does not like to download apps, even the most basic smartphone cameras can be powerful sensemaking tools when used creatively. Imagine you are teaching a biology class focused on studying plant growth. Instead of relying on textbook diagrams or premade videos, students could use their phone cameras to document the growth of seedlings over time. It's as simple as setting up a small plant station in a corner of the classroom and having students take a quick photo each day before class starts. Now, here's where the magic of technology comes in. Most phones these days have built-in time-lapse features. There are also free apps that can stitch these daily photos into a time-lapse video. Suddenly, a process that seemed painstakingly slow transforms into a dramatic display of life in motion. Students can watch weeks of growth unfold in seconds, observing patterns that would be impossible to see in real time. As they watch their time-lapse creations, students might notice some surprising things. Maybe they see that their plants seem to grow more at night than during the day. Or perhaps they observe leaves slowly shifting position throughout the day, reaching for the sunlight—like nature's own solar panels. These observations can spark all sorts of questions: Why do plants grow more at night? How do leaves know where the sun is? What would happen if we changed the light conditions?

But let's not stop at Earth-bound subjects. Those same smartphone cameras can turn our eyes to the sky, opening up a whole new realm of celestial observations. Imagine students tasked with photographing the moon every night for a month. As they compile their images, they'll see the moon's phases change night by night, transforming from a thin crescent to a full moon and back again. This visual record can lead to rich discussions about the moon's orbit, its relationship with the Earth and sun, and why we see different phases from our vantage point on Earth. Or consider tracking the sun's position at the same time each day throughout the year. With careful observation and safe photography techniques (never looking directly at the sun!), students might notice that the sun's position in the sky changes over time. This can lead to fascinating explorations of Earth's tilt, the reasons for seasons, and how our planet's movement affects life on Earth. It's a perfect blend of astronomy and Earth science, all captured with the camera in their pockets.

The beauty of this approach is that it turns every student into a researcher. They're not just passively receiving information about plant growth or celestial movements. Rather, students are actively documenting and analyzing these phenomena. And because they're using their own familiar devices, the barrier to entry is incredibly low. You don't need expensive equipment or a high-tech lab—just some seeds, soil, and the camera that is already in every student's pocket. These simple phone camera exercises can lead to deeper questions about biology, astronomy, and Earth science, in turn setting the stage for more advanced investigations. Students might want to experiment with different light conditions for their plants, track moon phases against tidal changes, or correlate the sun's position with local temperature variations. All of these investigations can be documented and analyzed using the same accessible technology, bridging the gap between everyday observations and complex scientific concepts.

Online simulation tools, many of which are free and accessible via web browsers, offer another avenue for sensemaking. For example, PhET simulations enable students to manipulate variables in complex systems and immediately see the results. Whether it's adjusting the concentration of solutions in a chemistry simulation or changing the mass of objects in a physics simulation, these tools allow for rapid experimentation and hypothesis testing. This quick feedback loop is invaluable for sensemaking, as it allows students to quickly test their ideas and refine their understanding.

Even social media platforms can be leveraged for sensemaking in STEM. Students can use platforms like Instagram, YouTube, or TikTok to document and share their scientific observations or experiments. This isn't just about making science "fun" or "relatable," although that can be a nice side effect. It's about encouraging students to communicate their scientific thinking clearly and concisely, a crucial skill in the sensemaking process. When students have to explain a concept or a finding in a short video, they're forced to distill their understanding to its essence, which often leads to deeper comprehension. Students can also find related phenomena that can be very helpful in the sensemaking process when using certain social media platforms! Almost weekly, I have students approach me to show me a video that they believe is connected to a phenomenon we are trying to figure out in class (or a phenomenon we recently figured out). What could be better than that? Rather than just sharing memes, students are sharing phenomena and, in doing so, even providing a little formative assessment opportunity as you ask them how this new video relates to the initial phenomenon!

It's important to note that the goal of using these technologies is not to replace traditional scientific methods or thinking. Rather, the goal is to enhance and support the sensemaking process. While the use of technology provides many new ways to observe, measure, analyze, and communicate, the human capacities of critical thinking and scientific reasoning still come from the students. A slow-motion video of a falling object doesn't explain gravity, but it provides rich data that students can use to construct their own understanding of gravitational effects.

CHAPTER 4

Assessing Sensemaking

Over the course of this book, we've explored how sensemaking in STEM supports student learning. At this point, you might be wondering, *How do we know what students are actually learning from all this*? It's a great question, and it gets to the heart of why we do what we do as educators. After all, cool experiments and engaging activities are fantastic, but our ultimate goal is student learning, right? When it comes to sensemaking, traditional assessment methods often fall short. Multiple-choice tests or end-of-unit exams might tell us what facts students can recall, but they don't necessarily reveal the depth of understanding or the thinking processes that sensemaking aims to develop. So how can we gather evidence of learning that truly reflects the rich, complex thinking of sensemaking we encourage our students to exercise when they are authentically engaged and actively participating in the learning process?

The key is not to think of assessment as a separate activity that happens after the learning process. Instead, we should think of assessment as an integral part of the learning process itself. In other words, we need to build assessment practices into our instruction, doing so in ways that feel natural and meaningful to students. Let's explore some approaches that both teachers and instructional leaders can use to make assessment a more organic component of the teaching and learning process. One powerful strategy is to leverage the very technology we're using for sensemaking as a tool for assessment. Remember those time-lapse videos of plant growth we discussed in the last chapter? Why not ask students to narrate their videos, explaining what they observed and hypothesized about the reasons behind the patterns they saw? This isn't just a cool presentation. It's a rich assessment of their ability to make observations, identify patterns, and construct explanations based on evidence. As teachers, we can listen for the connections they make to prior knowledge, the questions they generate, and even the scientific vocabulary they choose to use and how. All of this gives us valuable insight into their thinking and learning.

Similarly, when students are engaging with a phenomenon, we can ask them to keep a digital journal of their observations. This could be as simple as a shared online document or as elaborate as a class blog. The key is to encourage students to not just record what they see, but to also reflect on it. What patterns did they notice? What surprised them? What questions did this reflective practice raise? By reading students' reflections regularly, we can gain a window into their evolving understanding and can adjust our teaching accordingly.

But assessment doesn't have to be a solitary activity. In fact, some of the most powerful assessments happen when students interact with each other. Consider the following: What if—instead of a pen-and-paper test—the class were to set up a gallery walk where student teams present their findings after they have collected data using their phone sensors (i.e., those accelerometer readings)? In creating this gallery walk, students would present their team's data on large sheets of paper or digital displays around the room. As students circulate and discuss each other's work, you (as educator/facilitator) would listen in on their conversations. Are students asking thoughtful questions? Are they challenging each other's interpretations? Are they suggesting alternative explanations? This type of peer-to-peer interaction reinforces learning while also giving you a chance to assess their ability to engage in scientific discourse.

For school instructional leaders looking to support teachers in this assessment approach—that is, assessment as part of the active learning process—consider organizing *sharing sessions* for teachers to collaborate around exploring their assessment process. Here, teachers can present examples of student work from their lesson-embedded assessments. Sharing sessions not only celebrate student thinking; they also help the teaching staff calibrate their understanding of what good sensemaking looks like at different grade levels. Looking at these kinds of artifacts of student reasoning will enable teachers to more clearly see how scientific knowledge builds in complexity over the trajectory of K–12 education.

We have also discussed the SEP Developing and Using Models numerous times throughout this book. This strategy works well for both learning and assessment purposes. When students create a model based on their current understanding of a phenomenon, it provides a window into what they know while also revealing the gaps in their current understanding. Whether the model is a diagram, a concept map, or even a stop-motion video built from simple materials, the goal is not perfection. Rather, the goal is to externalize student thinking at a key moment in the unit. Furthermore, modeling is not a one-time process or creation upon which to assess student learning, because students revise their models over time. On this point, both the students and teacher will be able to see how students' understanding has been evolving. It's assessment and learning rolled into one!

Remember, the primary goal of assessment isn't to generate a grade (although they certainly do inform the gradebook). The real power lies in using these insights to guide our teaching and to help students reflect on their own learning journeys. By integrating assessment seamlessly into our sensemaking activities, we create a classroom culture in which curiosity, reflection, and continuous improvement become the norm. For teachers new to this approach, start small. Pick one sensemaking activity and try incorporating one of these assessment strategies. Pay attention to what you learn about your students' thinking and how it influences your next instructional moves. For school instructional leaders, consider how you can create spaces and opportunities for teachers to share and discuss their embedded assessments. Could you dedicate part of a staff meeting to the analysis of student work based on sensemaking activities?

STEM NOTEBOOKS

In terms of student ownership of their learning, there is one tool that's been a game-changer in my work: STEM notebooks. You see, real work in STEM isn't a neat, linear process. It's full of false starts, revised hypotheses, and *aha!* moments that come when you least expect them. Professional scientists have long used notebooks to track this messy, iterative process. So why shouldn't our students do the same? One of the biggest challenges I've faced as an educator is helping students see how their understanding evolves over time. How many times have you heard a student say "I've always known that!" when you knew for a fact that they had a completely different idea just a few weeks ago? It's not that they're intentionally being dishonest. It has to do with the fact that, as humans, our brains are really good at updating our knowledge and forgetting that we ever thought differently. This is where science notebooks come in. By giving students a space to record their thoughts, questions, and observations over time, we're providing them with a concrete record of their learning journey. It's like creating a time capsule of their thinking that they can revisit and reflect on throughout the year and beyond!

Now, in terms of format, you might be wondering, *Which is better, digital or paper*? The answer is . . . it depends! Both have their strengths, and the best approach often involves using them in complementary ways. Digital notebooks, for instance, offer some exciting possibilities. The digital notebooks I often use are just Google Slides where the page has been rotated to look like a notebook. I create a cover page and a table of contents, and then set up a number of different types of pages so students have options for recording their observations and thoughts. Depending on the scenario, I create versions of STEM notebooks that include information from previous experiences both in our classrooms and from previous years of studying the same concept (i.e., elementary school, middle school, etc.). Consider use of these notebooks as reminders of student thinking that can take many forms and serve a number of purposes. If district teams of teachers plan to meet with different grade-level teams, these groups can use those meetings as opportunities to leverage the work students have done in the past! Imagine a notebook for a high school biology class that includes a number of models from the previous work of this class in their life science units from middle school. While this does take some coordination, including work from earlier in the year and before is a powerful way to demonstrate how learning builds over time. These cumulative notebooks can serve as reminders and tools for students to build on the previous work they have done and to kick-start their learning in high school.

Another big benefit of digital notebooks is that they are always accessible for teachers, parents, and anyone who has a link to them, which means that you can take a peek inside at any moment! And they allow students to capture their learning in multiple formats. For instance, a student exploring the concept of momentum using a slow-motion video can embed the actual video alongside their typed observations. They can also sketch out their initial ideas using a digital stylus or online drawing tool, then update those sketches as their understanding evolves. Students can link to simulations they've used or articles they have read. They can even create models that incorporate images from the web to help them represent their thinking in ways they might find difficult to draw in a pen-and-paper notebook. In essence, they're creating a multimedia record of their sensemaking process that has no limits. Digital notebooks also make it incredibly easy for students to share their thinking with others. With just a few clicks, a student can share a specific entry with a classmate for peer feedback, or share with you for a quick formative check. This ability to quickly go public with their reasoning encourages students to refine their ideas and pushes them to be clearer in their explanations.

At the same time, let's not discount the value of traditional paper notebooks. There is something tangible and immediate about putting pen to paper that can't be replicated digitally. Paper notebooks are fantastic for quick sketches, for jotting down sudden insights, or for use in environments where digital devices might be impractical or at risk. When I introduce my concept of "notebooking," I often bring to class my notebooks from the years when I was doing animal behavior research using my Rite in the Rain notebooks. These notebooks blew my mind when I first was introduced to them. I can write in this thing even when it is raining? With a pencil and this paper notebook, I always had a place to capture observations and questions or record data about something that I was trying to figure out. I spent time in England, Ireland, and Scotland doing songbird vocalization research, and my Rite in the Rain notebooks were lifesavers during weeks when I saw more rainy days than sunny ones. I spent a lot of time observing great spotted woodpeckers in northern England, dedicating entire days to observing nesting behaviors of males and females in order to see if there were differences in their parental care for their young. When I show these notebooks from the past to my students, I make sure to point out that my notebooks often started out messy and became more organized as I figured out what I was trying to explain on the pages. I also highlight for my students the fact that there are more questions than answers in these notebooks, while also making note of whole pages of writing, whole pages of drawings, whole pages of data, and every combination of the sort. Using this kind of paper notebook is an opportunity to capture what you are thinking in any format that works best for you. For instance, John Muir Laws, a naturalist, artist, and educator, has a ton of resources available on his website (https://johnmuirlaws.com) to support students in creating nature journals. The site, and his accompanying resources, are worth checking out because you can download and use examples that will help students get started in recording and drawing their observations and ideas.

The tactile nature of paper notebooks can also aid memory and understanding for many students. The act of physically writing has been shown to enhance retention in ways that typing sometimes does not. Plus, flipping through pages of a physical notebook can provide a different kind of overview of one's learning journey, in contrast to scrolling through a digital document. In my classroom, I've found that a hybrid approach often works best. We might use paper notebooks for field observations or quick in-class reflections, then transfer and expand on these ideas in our digital notebooks. Or we might use digital tools to collect and analyze data, but use paper notebooks to sketch out our initial models and hypotheses. The key is to help students understand the value of both approaches and to give them the flexibility to choose the best tool for each situation. By doing so, we're not just teaching science content—we're helping students understand how scientific thinking actually works in various contexts.

DISCOURSE CIRCLES AND STUDENT-DRIVEN DISCUSSIONS

One method of assessment that might seem a bit unconventional at first glance, but which has become an invaluable tool in my work, is the Discourse Circle. Don't let the fancy name fool you. At its core, the Discourse Circle is simply a structured way of getting students talking about their ideas. But trust me, it's so much more than just a traditional class discussion. Instead of the typical classroom setup with neat rows all facing the front, your students are seated in a circle. Everyone can see one another, and the setting itself sets the stage for establishing a focus on the sharing of ideas. But the magic isn't in the seating arrangement—it's in what happens next. This isn't a free-for-all discussion or a simple sharing of what happened in class that day. Instead, it's a carefully orchestrated conversation about claims, evidence, and reasoning related to the specific concept you're exploring.

In Discourse Circles, your role as the teacher shifts from *sage on the stage* to *guide on the side*. I have been lucky enough to work extensively with Sarah Michaels, a sociolinguist and professor emerita from Clark University, who has helped me to understand the potential of these discussions and how important it is to meticulously plan for their success. When these discussions are done well, those leading the discussion have taken the time to think through how the discussion will unfold, crafting questions and prompts that will push students to dig deeper into their understanding, challenge one another's ideas, and back up their claims with evidence. It's like setting up dominoes. You arrange everything just so and then watch as the conversation unfolds—with each student's contribution triggering the next. Now, you might be thinking, *This sounds great, but how is this an assessment*? Well, that's where the real beauty of Discourse Circles comes in. As the conversation flows, you're not just a passive observer. You're actively tracking who's saying what by noting the claims students make, the evidence they cite, and the reasoning they use to connect the claims with the evidence. In carrying out this kind of assessment, some teachers I know use a simple diagram, jotting down student names and drawing lines to show who responds to whom. Others use a more detailed chart to track specific scientific ideas as they emerge in the discussion. This "discussion map" becomes a powerful assessment tool. At a glance, you can see which students are actively participating, who's making connections between ideas, and who's challenging claims with counterevidence. You start to get a clear picture of not just what students know, but also of how they're thinking about scientific concepts.

That said, the value of the Discourse Circle goes even deeper than that. As students engage in these discussions, they're not just regurgitating facts. They're actually engaging in the practices of science, meaning that they are arguing from evidence, constructing explanations, and critiquing others' reasoning. They're applying crosscutting concepts as they look for patterns or as they discuss cause-and-effect relationships. And, of course, they're grappling with core disciplinary ideas. In other words, the Discourse Circle becomes a window into students' progress on all three dimensions of the NGSS. Let me share a quick example from my own classroom. We were studying ecosystems, and I set up a Discourse Circle to discuss the impact of introducing a new species into an established ecosystem. As the discussion unfolded, I watched students move from simplistic cause-and-effect statements to more nuanced discussions of energy flow and interdependence. One student made a claim about competition for resources. Another countered with evidence about niche partitioning from our recent lab work. A third student jumped in to connect these ideas to the stability of ecosystems—a core disciplinary idea we had been exploring. As I mapped this discussion, I could see in real time how students were connecting ideas. I also made note of which ideas were only partially formed and kept track of which students still needed more time/evidence to add to their conceptual model of ecosystem dynamics. This mapping process enabled me to easily identify students who were ready for more challenging content. Overall, I came to see this kind of rich, qualitative data as gold for informing my instruction moving forward.

Now, I know what some of you might be thinking: *This sounds great in theory, but how do I actually make it work in my classroom*? It's a valid question. The Discourse Circle does not just happen. It takes careful planning and practice to design and implement this process. Start small. Maybe begin with a simple claim-evidence-reasoning framework for discussions. As students (and you!) get more comfortable, you can introduce more complex discussion protocols. The Discourse Circle is also a great place to work with your talk moves. When I was new to facilitating these discussions, I would always keep a laminated copy of the talk moves checklist on my clipboard so that I could track which talk moves I used and when. You can also use sentence starters with younger learners, being strategic about what to include. For example, if I wanted to focus on helping students agree or disagree with their classmates using evidence, I would have sentence strips positioned on the floor in the center of our circle. The sentence strips said things like, "I agree with [name] because our group found . . ." or "I disagree with [name] because we observed . . .".

For those of you in instructional leadership roles, this is where you can really support your teachers. Consider organizing professional development sessions in which teachers can practice facilitating these discussions with one another. These rehearsals can be extremely valuable when a teacher is planning for a specific discussion they are about to lead in their classroom. A professional development session can provide a supported space for teachers to carefully think through how they want the planned discussion to play out. As a follow-up, another professional development session could focus on the results of the actual student-centered discussions. As instructional leaders, you could create this session to serve as a space in which teachers would share their discussion maps and collaboratively analyze what the maps revealed about student thinking. In fact, it can be extremely helpful to see a discussion map showing that only a few students (or teachers in this case) were talking in a Discourse Circle activity; then take a step back and reflect on the norms of equitable and respectful talk focused on reasoning. Ultimately, these maps are tangible representations of what happened during a classroom discussion. My experience has been that students and teachers find discussion maps to accurately reflect the dynamics of their classroom communities. Finally, as an instructional leader, you might even consider doing learning walks focused specifically on student discourse in STEM classrooms, even offering to map the discussion for your teachers so as to support their norms.

One major key to making Discourse Circles successful is to integrate them into your classroom norms from the get-go. Students need to understand that these aren't just fun discussion days, but a core part of how science learning happens in your classroom. They need to feel safe sharing their ideas, challenging each other respectfully, and, yes, even being wrong sometimes. Building this culture takes time, but the payoff in terms of student learning is immense. As you start to use Discourse Circles more regularly, you'll find that they become a powerful tool not just for assessment, but for instruction as well. The insights you gain from these discussions will help you tailor your teaching to your students' needs, raise new ideas that might not have emerged before, build on emerging understandings, and push students to ever-deeper levels of scientific thinking.

EXAMPLE RUBRICS FOR ASSESSING SENSEMAKING

As we've explored throughout this chapter, assessing sensemaking in STEM education requires a nuanced approach that goes beyond traditional testing methods. To help you put these ideas into practice, let's dive into some concrete examples of rubrics designed specifically for evaluating sensemaking. These rubrics aren't just about checking boxes or assigning grades. Rather, they are tools to help you and your students understand the depth and quality of the scientific thinking happening in your classroom.

We'll start with a one-point rubric focused on scientific reasoning. This rubric provides a flexible framework for identifying areas of strength and improvement across various scientific practices. Then we'll examine more detailed rubrics that break down specific aspects of sensemaking—from engaging with phenomena to communicating scientific ideas. We'll also look at a three-dimensional rubric that integrates science and engineering practices, crosscutting concepts, and disciplinary core ideas, aligning closely with the NGSS framework. Next, we'll explore rubrics that can be used for (a) assessing discussions focused on reasoning, (b) assessing STEM notebooking, and (c) assessing technology integration. Finally, we'll focus on sensemaking in mathematics and sensemaking in engineering, as these two areas have slightly different approaches with important distinctions to highlight.

But remember, these rubrics aren't meant to be one-size-fits-all solutions. They're starting points—designed to be adapted and refined based on your specific classroom needs, grade level, and the particular phenomena or problems your students are exploring. As you review these examples, think about how you might modify them to better suit your teaching context or to focus on particular aspects of sensemaking you want to emphasize. The goal here isn't just assessment for assessment's sake. These rubrics are designed to provide meaningful feedback to students, guide your instructional decisions, and ultimately deepen the sensemaking process in your STEM classroom.

Rubric—One-Point Rubric: Scientific Reasoning Assessment

The one-point rubric format offers a powerful tool for assessing scientific reasoning while avoiding the rigid constraints of traditional scoring systems. Unlike conventional rubrics that might lock us into predetermined levels, this format creates space for recognizing unique aspects of student thinking and provides specific, actionable feedback. The center column establishes clear expectations for proficient performance, while the outer columns allow us to document both areas where students might need additional support and moments where they demonstrate exceptional scientific reasoning. This flexibility is particularly valuable when assessing sensemaking, as students often surprise us with novel approaches and unexpected insights that might not fit neatly into traditional scoring bands.

Areas for Improvement	Criteria for Proficient Scientific Reasoning	Areas of Excellence
	1. Asking Questions and Defining Problems: Student formulates relevant scientific questions or clearly defines problems based on observations or prior knowledge.	

Areas for Improvement	Criteria for Proficient Scientific Reasoning	Areas of Excellence
	2. Developing and Using Models: Student creates or uses models to explain phenomena, make predictions, or test ideas.	
	3. Planning and Carrying Out Investigations: Student designs logical steps for scientific investigations, considering variables and controls.	
	4. Analyzing and Interpreting Data: Student organizes data, identifies patterns, and draws valid conclusions based on evidence.	
	5. Using Mathematics and Computational Thinking: Student applies mathematical concepts or computational methods to analyze data or solve problems.	
	6. Constructing Explanations: Student constructs explanations based on valid and reliable evidence obtained from various sources.	
	7. Engaging in Argument From Evidence: Student uses reasoning and evidence to support claims, evaluate others' ideas, and revise their own thinking.	
	8. Obtaining, Evaluating, and Communicating Information: Student critically assesses scientific information from various sources and effectively communicates scientific ideas.	

Tips for Using This Rubric

1. Use specific examples that capture students' sensemaking processes. Instead of writing "needs more detail," note "could strengthen explanation by connecting their initial ideas about plant growth and their connection to the light intensity data."
2. Share the rubric with students before assignments and encourage them to self-assess not just their final products, but their thinking processes and how they made sense of scientific phenomena.
3. Consider using different-colored highlighters or digital tags to track how students' sensemaking evolves over time, particularly how their explanations and models become more sophisticated.

4. Document unexpected or creative approaches in the "Areas of Excellence" column—sensemaking often leads to novel ways of understanding phenomena that we might not have anticipated.
5. Use the rubric to inform your instruction by identifying which aspects of sensemaking (like developing models or constructing explanations) need additional support across your class.
6. Collect evidence across multiple contexts and activities, as sensemaking develops gradually as students encounter new phenomena and build connections between concepts.

Rubric—SEP 1: Engaging With Phenomena and Asking Questions

When it comes to sensemaking, everything starts with how students initially engage with phenomena and the questions they generate. This rubric focuses on those crucial first moments of scientific inquiry—the spark that ignites deeper investigation. By assessing how students observe, question, and connect to their prior knowledge, we can better understand their entry points into scientific thinking and guide them toward more sophisticated engagement with phenomena. The progression from beginning to advanced levels isn't just about doing more—it's about seeing more deeply, questioning more thoughtfully, and making richer connections.

Criteria	Beginning (1)	Developing (2)	Proficient (3)	Advanced (4)
Observation of phenomena	Makes limited observations of the phenomenon	Makes basic observations of the phenomenon	Makes detailed observations of the phenomenon	Makes comprehensive observations, noticing subtle aspects of the phenomenon
Question Formulation	Asks simple or unrelated questions	Asks relevant but surface-level questions	Asks probing questions that relate to the phenomenon	Asks insightful questions that challenge existing understanding or propose new investigations
Connection to Prior Knowledge	Shows little connection to prior knowledge	Makes basic connections to prior knowledge	Clearly connects the phenomenon to prior knowledge	Synthesizes prior knowledge with new observations to form hypotheses

Tips for Using This Rubric

1. Listen for the unexpected—students often ask questions that reveal unique ways of making sense of phenomena that we as teachers might not have anticipated.
2. Document students' initial questions verbatim, as the precise wording reveals their sensemaking process and what aspects of the phenomenon they're trying to understand.
3. When students connect questions to prior knowledge, note how they're using both scientific ideas and personal experiences to generate new lines of inquiry.
4. Use this rubric early in a unit to identify which students might need support in developing deeper questions, and which might be ready to pursue more complex aspects of the phenomenon.
5. Track how students' questions evolve as they engage with a phenomenon multiple times—notice how their questions become more sophisticated as their understanding develops.

Rubric—SEP 2: Developing and Using Models

Models are powerful tools for sensemaking—they help students externalize their thinking and wrestle with complex phenomena in tangible ways. This rubric focuses on how students develop, use, and revise models as part of their sensemaking over time. Whether students are drawing diagrams, building physical representations, or creating digital simulations, their models provide windows into how they're making sense of scientific phenomena. The progression from beginning to advanced levels reflects students' growing ability to use models not just as representations, but as thinking tools that evolve alongside their understanding.

Criteria	Beginning (1)	Developing (2)	Proficient (3)	Advanced (4)
Model Creation	Creates simplistic models with limited relevance	Creates basic models that partially represent the phenomenon	Creates clear models that accurately represent key aspects of the phenomenon	Creates sophisticated models that represent complex relationships within the phenomenon
Model Usage	Uses models in a limited way to describe the phenomenon	Uses models to describe and partially explain the phenomenon	Effectively uses models to explain the phenomenon and make predictions	Uses models to explain the phenomenon, make predictions, and generate new questions or hypotheses

Criteria	Beginning (1)	Developing (2)	Proficient (3)	Advanced (4)
Model Revision	Shows resistance to revising models	Makes minor revisions to models when prompted	Willingly revises models based on new evidence or understanding	Proactively refines and revises models, explaining the reasons for changes

Tips for Using This Rubric

1. Look for how students represent both observable and unobservable elements—sophisticated models often include invisible mechanisms to explain visible phenomena.
2. Remind students that models are thinking tools—encourage them to include *all* their ideas, even competing explanations or uncertainties. A "messy" model showing deep thinking is more valuable than a neat but superficial one.
3. Watch how students use their models to reason—do they use them to tell causal stories rather than just describe what they see?
4. Notice whether students use their models as evidence to support claims or address counterarguments.
5. Pay attention to how students use their models to show change over time—before, during, and after key events.
6. Document moments when students identify the need to revise their models—these often reveal deepening understanding.
7. Listen for students' reasoning about why they're making changes to their models—what new evidence or insights are driving these revisions?

Rubric—SEP 3: Planning and Carrying Out Investigations

The heart of scientific sensemaking often lies in how students design and conduct investigations to answer their questions about phenomena. This rubric focuses on the path from initial investigation design through data collection, all while maintaining appropriate safety and ethical considerations. Rather than following cookbook-style procedures, we want to see students thinking critically about what evidence they need to collect and how best to collect it. The progression from beginning to advanced levels reflects growing sophistication in students' ability to plan investigations that will help them make sense of scientific phenomena.

Criteria	Beginning (1)	Developing (2)	Proficient (3)	Advanced (4)
Investigation Design	Struggles to design a relevant investigation	Designs a basic investigation that partially addresses the question	Designs a clear, relevant investigation that addresses the question	Designs a sophisticated investigation that thoroughly addresses the question and considers potential variables
Data Collection	Collects data inconsistently or inaccurately	Collects relevant data, but may have gaps or inconsistencies	Collects accurate and complete data relevant to the investigation	Collects comprehensive data, including multiple trials or conditions where appropriate
Safety and Ethics	Shows limited awareness of safety and ethical considerations	Demonstrates basic awareness of safety and ethical considerations	Consistently follows safety procedures and considers ethical implications	Proactively addresses safety and ethical considerations, including potential impacts of the investigation

Tips for Using This Rubric

1. Listen for how students connect their investigation design to their questions—are they collecting evidence that will actually help them figure out the phenomenon they're trying to understand?
2. Notice whether students can explain why they chose specific variables to measure—this reveals their thinking about what might be important in the phenomenon.
3. Watch for students' real-time adjustments to their data collection methods—these moments often show deepening understanding of what evidence they believe they need to prove or disprove their reasoning.
4. Pay attention to how students plan to document their investigation—their choices about what data to record and how often reveal their thinking about what matters in the phenomenon.
5. Look for students who anticipate potential problems or sources of error in their investigation plans—this shows sophisticated thinking about the quality of evidence they'll need.
6. Document how students justify the number of trials or measurements they plan to make—this reveals their understanding of reliability in scientific investigations.

Rubric—SEP 4: Analyzing and Interpreting Data

Making sense of data is more than just crunching numbers or making graphs—it's about finding patterns and stories that help us understand phenomena more deeply. This rubric focuses on how students work with data as a tool for sensemaking, from initial organization through analysis and interpretation. We're looking for students to move beyond simply plotting points or calculating averages to really wrestling with what their data is telling them about the phenomenon they're investigating. The progression from beginning to advanced levels reflects students' growing ability to use data as evidence for their scientific thinking.

Criteria	Beginning (1)	Developing (2)	Proficient (3)	Advanced (4)
Data Organization	Struggles to organize data in a meaningful way	Organizes data in a basic format, but may lack clarity	Organizes data clearly and effectively	Organizes data in multiple ways to reveal different aspects of the findings
Data Analysis	Performs limited or incorrect analysis of data	Performs basic analysis of data with some accuracy	Accurately analyzes data using appropriate methods	Performs comprehensive analysis, including statistical tests where appropriate
Interpretation	Draws limited or incorrect conclusions from data	Draws basic conclusions that partially reflect the data	Draws logical conclusions that accurately reflect the data	Draws insightful conclusions, considering limitations and implications of the data

Tips for Using This Rubric

1. Watch for how students choose to organize their data—their organizational choices often reveal what patterns they think might be important in making sense of the phenomenon.
2. Notice when students try different ways of looking at their data—these moments of reorganization often show they're searching for deeper patterns or relationships.
3. Listen for students questioning their own data—are they considering what might be missing or what could explain unexpected results?
4. Pay attention to how students connect their data analysis back to their original questions about the phenomenon—are they using the data to build understanding?

5. Look for students who can explain *why* they chose specific ways to analyze their data—this reveals their thinking about what evidence they need to support their explanations.
6. Document how students use their data to support or challenge their initial ideas about the phenomenon they're investigating.

Rubric—SEP 5: Using Mathematics and Computational Thinking

Mathematics and computation aren't just about getting right answers—they're powerful tools for making sense of scientific phenomena. This rubric focuses on how students use mathematical representations, computational tools, and quantitative analysis to deepen their understanding of scientific concepts. Whether students are creating mathematical models, using spreadsheets to analyze data, or running simulations, we're looking for how these tools help them uncover patterns and relationships they might not otherwise see. The progression from beginning to advanced levels reflects students' growing ability to use mathematical and computational thinking to reveal deeper insights about phenomena.

Criteria	Beginning (1)	Developing (2)	Proficient (3)	Advanced (4)
Mathematical Representation	Uses limited or inappropriate mathematical representations	Uses basic mathematical representations with some relevance	Uses appropriate mathematical representations to describe phenomena	Creates sophisticated mathematical models that accurately represent complex phenomena
Computational Tools	Shows limited ability to use computational tools	Uses basic computational tools with some effectiveness	Effectively uses computational tools to analyze data or model phenomena	Leverages advanced computational tools to analyze complex data or create detailed simulations
Quantitative Analysis	Performs limited quantitative analysis	Performs basic quantitative analysis with some accuracy	Accurately performs quantitative analysis to support scientific arguments	Integrates advanced quantitative analysis to reveal subtle patterns or relationships in data

Tips for Using This Rubric

1. Watch for how students identify and represent key variables in phenomena—do they recognize which quantities matter and how they might be related?
2. Notice when students use mathematics to make predictions about phenomena—are they using equations or graphs to forecast what might happen under different conditions?
3. Look for students using digital tools to organize and analyze data, especially when dealing with large datasets where patterns might not be immediately visible.
4. Pay attention to how students test and refine their computational models—do they compare their model's predictions with actual observations?
5. Document how students use mathematical patterns they discover to support scientific arguments about cause and effect relationships.

Rubric—SEP 6 & 7: Constructing Explanations and Engaging in Argument From Evidence

Constructing explanations and engaging in argument from evidence are two closely related practices that reveal how students make sense of scientific phenomena. Whether students are explaining how something works or building a case for why something happens, they're using evidence and reasoning to develop their understanding. This rubric helps track how students progress from making simple claims to developing evidence-based explanations and engaging in scientific argumentation that deepens their understanding of phenomena.

Criteria	Beginning (1)	Developing (2)	Proficient (3)	Advanced (4)
Evidence Usage	Uses little or no evidence in explanations	Uses some evidence in explanations, but may be irrelevant or misinterpreted	Uses relevant evidence to support explanations	Synthesizes multiple sources of evidence to create comprehensive explanations
Scientific Reasoning	Shows limited scientific reasoning in explanations	Demonstrates basic scientific reasoning in explanations	Applies sound scientific reasoning to connect evidence and claims	Demonstrates advanced scientific reasoning, considering alternative explanations and limitations of evidence

Criteria	Beginning (1)	Developing (2)	Proficient (3)	Advanced (4)
Argumentation	Presents claims with little justification	Presents claims with some justification, but may struggle to address counterarguments	Engages in effective argumentation, addressing counterarguments	Engages in sophisticated argumentation, evaluating the strength of different claims and proposing new investigations to resolve uncertainties

Tips for Using This Rubric

1. Listen for how students connect their evidence to their claims—do they explain why specific evidence helps them make sense of the phenomenon?
2. Notice when students revise their explanations after encountering new evidence—these moments often reveal deeper sensemaking.
3. Watch for students challenging their own thinking—do they consider alternative explanations or identify gaps in their evidence?
4. Pay attention to how students respond to counterarguments—do they use evidence to strengthen their explanations or acknowledge limitations?
5. Look for students who distinguish between what their evidence directly shows and what they're inferring—this reveals sophisticated thinking about the strength of scientific arguments.
6. Document how students use scientific principles to explain why their evidence supports their claims, not just state that it does.
7. Notice when students propose new questions or investigations to strengthen their arguments—this shows they understand that scientific sensemaking is an ongoing process.

Rubric—SEP 8: Obtaining, Evaluating, and Communicating Information

Making sense of phenomena and solving problems isn't just about finding facts—it's about evaluating sources, synthesizing ideas, and sharing understanding effectively with others. And let's be honest, in a world where students are just as likely to learn about scientific phenomena from TikTok as textbooks (yes, we've all seen those viral "science experiment" videos), evaluating sources is more important than ever! This rubric helps track how students progress from simply gathering information to becoming skilled evaluators and communicators of scientific phenomena. Whether students are researching real-world problems, reading scientific explanations of phenomena, or sharing their own findings (hopefully with more academic rigor than their favorite influencer), we're looking at how they develop their ability to work with scientific information in meaningful ways.

Criteria	Beginning (1)	Developing (2)	Proficient (3)	Advanced (4)
Information Gathering	Gathers limited or irrelevant scientific information	Gathers basic scientific information from provided sources	Effectively gathers relevant scientific information from multiple sources	Comprehensively gathers and synthesizes information from a wide range of credible sources
Source Evaluation	Shows limited ability to evaluate source credibility	Demonstrates basic ability to evaluate source credibility	Effectively evaluates the credibility and bias of scientific sources	Critically evaluates sources, considering multiple perspectives and potential conflicts of interest
Scientific Communication	Communicates scientific ideas with limited clarity	Communicates scientific ideas with basic clarity; attempts to use some scientific terminology	Communicates scientific ideas clearly and accurately, using appropriate scientific terminology	Articulates complex scientific ideas with precision; adapts communication style effectively for different audiences

Tips for Using This Rubric

1. Listen for how students explain which sources helped them understand the phenomenon—their reasoning reveals how they're making sense of what counts as reliable scientific information.
2. Watch for students connecting information from different sources to explain a phenomenon—these synthesis moments often show deeper sensemaking.
3. Notice when students question how sources explain phenomena—do they consider the evidence behind different explanations?
4. Pay attention to how students communicate their understanding of phenomena to different audiences while maintaining scientific accuracy.
5. Look for students who can explain complex phenomena using both everyday language and scientific terminology appropriately.
6. Document how students use evidence from their sources to support their explanations of phenomena or proposed solutions to problems.

7. Notice when students identify gaps or contradictions in how different sources explain phenomena—this shows they're actively making sense rather than just collecting information.

Rubric—Three-Dimensional Rubric: Assessing Sensemaking Across SEP, CCC, and DCI

The next rubric provides an example of how to assess sensemaking across the three dimensions of the NGSS: science and engineering practices (SEPs), crosscutting concepts (CCCs), and disciplinary core ideas (DCIs). The example used here is based on a middle school (MS) performance expectation related to ecosystems.

MS-LS2-4: Construct an Argument Supported by Empirical Evidence That Changes to Physical or Biological Components of an Ecosystem Affect Populations.

Dimension	Beginning (1)	Developing (2)	Proficient (3)	Advanced (4)
Science and Engineering Practice: Engaging in Argument from Evidence	Makes claims about ecosystem changes with little or no supporting evidence	Presents basic arguments about ecosystem changes with some supporting evidence, but may struggle to connect evidence to claims	Constructs clear arguments about ecosystem changes, supported by relevant empirical evidence	Constructs sophisticated arguments about ecosystem changes, synthesizing multiple lines of empirical evidence and addressing potential counterarguments
Crosscutting Concept: Cause and Effect	Identifies simple cause-and-effect relationships in ecosystems without explaining mechanisms	Describes basic cause-and-effect relationships in ecosystems, with limited explanation of mechanisms	Explains cause-and-effect relationships in ecosystems, including some explanation of underlying mechanisms	Analyzes complex cause-and-effect relationships in ecosystems, explaining detailed mechanisms and considering multiple interacting factors
Disciplinary Core Idea: LS2.C: Ecosystem Dynamics, Functioning, and Resilience	Demonstrates limited understanding of how ecosystem changes affect populations	Shows basic understanding of how some ecosystem changes can affect populations	Explains how various physical and biological changes in an ecosystem can affect populations	Provides in-depth explanations of how diverse ecosystem changes affect populations, including indirect effects and feedback loops

Dimension	Beginning (1)	Developing (2)	Proficient (3)	Advanced (4)
Integration of Dimensions	Treats dimensions separately, with little connection between them	Makes basic connections between two dimensions, but may struggle to integrate all three	Effectively integrates all three dimensions to explain ecosystem changes and their effects	Seamlessly integrates all three dimensions, demonstrating a sophisticated understanding of how they interrelate in explaining ecosystem phenomena

Tips for Using This Rubric

1. **SEP (Engaging in Argument from Evidence):** Assess how well students construct and support their arguments about ecosystem changes.
2. **CCC (Cause and Effect):** Evaluate students' ability to identify and explain cause-and-effect relationships within ecosystems.
3. **DCI (Ecosystem Dynamics, Functioning, and Resilience):** Gauge students' understanding of how changes in ecosystem components affect populations.
4. **Integration:** Assess how well students bring together all three dimensions in their sensemaking process.

Rubric—Assessing Sensemaking in Scientific Discussions

Scientific sensemaking isn't just something that happens in students' heads—it comes alive in classroom discussions where ideas can be tested, challenged, and refined through interaction with peers. This rubric recognizes that powerful learning happens when students engage in productive scientific talk, whether they're proposing explanations for phenomena, evaluating evidence together, or building on each other's ideas. We're looking at multiple dimensions of discussion-based sensemaking: how students participate equitably, use evidence, employ talk moves, demonstrate reasoning, build on peers' contributions, and show their understanding of the phenomena or problems being discussed.

It's important to acknowledge that while scientific discussions might start with a single student sharing a tentative idea (you know the one—there's always that eager hand that shoots up before you've even finished the question), the use of thoughtful facilitation and productive discourse moves can help to evolve these into rich conversations where students collectively make sense of complex phenomena.

Criteria	Beginning (1)	Developing (2)	Proficient (3)	Advanced (4)
Equitable Participation	Participates minimally or dominates the discussion, showing little awareness of others' contributions	Participates in the discussion, but may not consistently engage with or acknowledge others' ideas	Participates actively while also making space for others, demonstrating respect for diverse perspectives	Actively promotes equitable participation by encouraging and amplifying others' voices, especially those less frequently heard
Use of Evidence	Rarely refers to evidence or uses irrelevant evidence to support claims	Sometimes uses evidence to support claims, but may misinterpret or oversimplify the data	Consistently uses relevant evidence to support claims, accurately interpreting data	Synthesizes multiple sources of evidence to support claims, addressing limitations and uncertainties in the data
Application of Talk Moves	Shows limited use of productive talk moves	Uses basic talk moves (e.g., "I agree/disagree because . . ."), but may not always use them effectively	Effectively uses a variety of talk moves to clarify, build on, or respectfully challenge ideas	Skillfully employs a wide range of talk moves to deepen the discussion, promote collaborative sensemaking, and encourage higher-order thinking
Focus on Reasoning	Shares ideas with little explanation of reasoning	Provides basic reasoning for ideas, but may struggle to connect concepts	Clearly explains reasoning, making logical connections between evidence and claims	Demonstrates advanced reasoning, considering alternative explanations and evaluating the strength of different arguments
Building on Others' Ideas	Rarely acknowledges or builds on others' contributions	Sometimes acknowledges others' ideas, but may not effectively integrate them into their own thinking	Regularly builds on and extends others' ideas, contributing to collective understanding	Consistently synthesizes and extends multiple perspectives, driving the discussion toward deeper collective sensemaking

Criteria	Beginning (1)	Developing (2)	Proficient (3)	Advanced (4)
Conceptual Understanding of Phenomena/ Problems	Demonstrates limited understanding of the core concepts related to the phenomenon or problem being discussed	Shows basic understanding of some key concepts, but may struggle to apply them to the phenomenon or problem	Demonstrates clear understanding of the core concepts and applies them accurately to explain the phenomenon or problem	Exhibits deep conceptual understanding, making insightful connections between concepts and the phenomenon or problem, considering multiple perspectives.

Tips for Using This Rubric

1. **Equitable Participation:** Look for students who not only contribute their own ideas, but who also create space for others to participate. This includes active listening by encouraging students who might not normally share and acknowledging diverse perspectives.
2. **Use of Evidence:** Assess how students incorporate data, observations, or other forms of evidence into their discussions. Look for accurate interpretation and appropriate application of evidence.
3. **Application of Talk Moves:** Evaluate students' use of productive discussion strategies such as:
   - Revoicing: "So, are you saying that . . . ?"
   - Asking for clarification: "Can you explain what you mean by . . . ?"
   - Challenging or countering: "But what about . . . ?"
   - Connecting ideas: "How does that relate to what [student] said earlier?"
4. **Focus on Reasoning:** Look for students who explain their thinking processes, not just their conclusions. Assess the logical coherence of their arguments.
5. **Building on Others' Ideas:** Observe how students respond to and incorporate their peers' contributions. This demonstrates collaborative sensemaking.
6. **Conceptual Understanding of Phenomena/Problems:** Evaluate how well students grasp and apply core scientific concepts to explain the phenomenon or problem under discussion. Look for the student's:
   - Accurate application of scientific principles.
   - Ability to explain phenomena at different scales or from multiple perspectives.

- Ability to make connections between different concepts or ideas.
- Ability to identify patterns or relationships within the phenomenon or problem.

Rubric—Assessing Sensemaking in STEM Notebooks

STEM notebooks aren't just places to record data or store worksheets—they're powerful tools where students can track their developing understanding of scientific phenomena. Think of them as windows into students' sensemaking journeys, where we can see their initial observations evolve into deeper scientific thinking. Each notebook becomes a personal record of how a student makes sense of scientific concepts and practices over time.

This rubric helps us look at how students use their notebooks as personal thinking spaces, from recording their first observations (which might start with "This is so cool!" or "I have no idea what I'm looking at . . .") to developing sophisticated explanations of phenomena. Whether students are sketching models, recording data, asking questions, or revising their thinking, these notebooks should capture the authentic process of trying to figure things out. While students will develop their own unique approaches to organizing and documenting their work, what matters most isn't the aesthetics—it's the evidence of students actively wrestling with scientific ideas and practices.

Criteria	Beginning (1)	Developing (2)	Proficient (3)	Advanced (4)
Observations and Data Collection	Records minimal or irrelevant observations; data collection is incomplete or disorganized	Records basic observations and data, but may lack detail or organization	Records detailed, relevant observations and data in an organized manner	Records comprehensive, precise observations and data, using appropriate methods (e.g., tables, graphs) to organize information effectively
Questions and Wonderings	Rarely records questions or wonderings about scientific phenomena	Occasionally records questions, but questions may be superficial or disconnected from observations	Regularly records relevant questions and wonderings that arise from observations and class activities	Consistently records insightful questions and wonderings that demonstrate curiosity and drive further investigation

Criteria	Beginning (1)	Developing (2)	Proficient (3)	Advanced (4)
Models and Explanations	Attempts to create models or explanations are minimal or unclear	Creates basic models or explanations that partially represent scientific concepts or phenomena	Develops clear models and explanations that accurately represent scientific concepts or phenomena	Creates sophisticated models and explanations that represent complex relationships and demonstrate deep understanding of concepts
Connections and Patterns	Makes few or no connections between ideas or identifies limited patterns	Makes some basic connections between ideas and identifies simple patterns	Consistently identifies meaningful connections and patterns across different concepts or phenomena	Synthesizes ideas to identify complex patterns and make insightful connections across multiple concepts or investigations
Reflections and Revisions	Shows little evidence of reflecting on learning or revising ideas	Occasionally reflects on learning or revising ideas, but may not explain reasoning for changes	Regularly reflects on learning and revising ideas, explaining reasons for changes in thinking	Consistently engages in thoughtful reflection, revising ideas with clear explanations and considering multiple perspectives
Use of Evidence	Rarely uses evidence to support claims or explanations	Sometimes uses evidence to support claims, but may be irrelevant or misinterpreted	Consistently uses relevant evidence to support claims and explanations	Synthesizes multiple sources of evidence to support claims, addressing limitations and uncertainties
Scientific Reasoning	Demonstrates limited scientific reasoning in entries	Shows basic scientific reasoning, but may struggle to connect evidence to claims	Applies sound scientific reasoning to connect evidence and claims	Demonstrates advanced scientific reasoning, considering alternative explanations and evaluating the strength of evidence
Integration of Scientific Practices	Shows limited use of scientific practices in notebook entries	Incorporates some scientific practices, but may not integrate them effectively	Effectively integrates multiple scientific practices in notebook entries	Seamlessly integrates a wide range of scientific practices, demonstrating a sophisticated approach to scientific inquiry

Tips for Using This Rubric

1. **Observations and Data Collection:** Look for detailed, accurate records of observations and data. Higher levels should show more sophisticated methods of data organization.
2. **Questions and Wonderings:** Evaluate the quality and relevance of questions students record. Higher levels should show questions that drive deeper inquiry.
3. **Models and Explanations:** Assess both the accuracy and complexity of students' models and explanations. Look for progression from simple representations to more nuanced, detailed models.
4. **Connections and Patterns:** Look for evidence that students are linking ideas across different investigations or concepts, and also identifying patterns in their observations or data.
5. **Reflections and Revisions:** Check for evidence of metacognition—students thinking about their own learning and how their ideas have changed over time.
6. **Use of Evidence:** Evaluate how students incorporate evidence into their explanations and claims. Higher levels should show more sophisticated use of multiple sources of evidence.
7. **Scientific Reasoning:** Assess the logic and depth of students' reasoning skills as they connect evidence to claims and explanations.
8. **Integration of Scientific Practices:** Look for evidence that students are engaging in multiple scientific practices (e.g., asking questions, planning investigations, analyzing data) in an integrated way.

Remember, this rubric can serve both formative and summative purposes. During units, use it formatively to provide feedback that guides students toward more effective use of their notebooks for sensemaking—helping them see their notebooks as thinking tools rather than just repositories for assignments. While notebooks can also be evaluated summatively at key points (like end of units), even then the focus should remain on how students use them to develop and refine their understanding of scientific concepts and practices. The goal isn't to create perfect notebooks, but to support students in documenting how their thinking evolves over time in meaningful (and often complex) ways.

Rubric—Assessing Sensemaking: Triple E Framework for Technology Integration

Technology in STEM classrooms isn't just about using the latest digital tools—it's about how these tools can enhance students' ability to make sense of scientific phenomena. The Triple E Framework (Kolb, 2011) helps us evaluate technology integration through three crucial lenses: Engagement, Enhancement, and Extension. This rubric looks at how technology can motivate deeper sensemaking, reduce cognitive load so students can focus on important scientific ideas, and connect classroom learning to real-world contexts. Whether students are using simulations, collecting data with sensors, or collaborating through digital platforms, we want to see technology supporting rather than distracting from the core work of making sense of STEM ideas.

Criteria	Beginning (1)	Developing (2)	Proficient (3)	Advanced (4)
		Engagement		
Motivation	Technology is used, but doesn't increase student motivation to engage in sensemaking activities	Technology use somewhat increases student motivation to engage in sensemaking activities	Technology clearly increases student motivation to engage in sensemaking activities	Technology significantly boosts student motivation, leading to deep engagement in sensemaking activities
Focus	Students are easily distracted by technology, hindering sensemaking processes	Students maintain focus on sensemaking, but with some technology-related distractions	Students maintain focus on sensemaking with minimal technology-related distractions	Technology use enhances student's focus, deepening their sensemaking processes
Social/ Emotional Learning	Technology use doesn't support social or emotional aspects of sensemaking	Technology somewhat supports social or emotional aspects of sensemaking	Technology effectively supports social and emotional aspects of sensemaking	Technology significantly enhances social and emotional aspects of sensemaking, fostering collaboration and empathy
		Enhancement		
Cognitive Load	Technology increases cognitive load, making sensemaking more difficult	Technology has a neutral effect on cognitive load during sensemaking	Technology reduces cognitive load, making sensemaking more accessible	Technology significantly reduces cognitive load, allowing for deeper and more complex sensemaking

Criteria	Beginning (1)	Developing (2)	Proficient (3)	Advanced (4)
Understanding	Technology use doesn't enhance understanding of scientific concepts	Technology somewhat enhances understanding of scientific concepts	Technology clearly enhances understanding of scientific concepts	Technology significantly deepens understanding of scientific concepts, revealing complex relationships
Scaffolding	Technology provides little to no scaffolding for sensemaking	Technology provides basic scaffolding for sensemaking	Technology provides effective scaffolding, supporting various levels of sensemaking	Technology provides adaptive scaffolding, optimally supporting each student's sensemaking process
		Extension		
Real-World Connection	Technology doesn't connect sensemaking to real-world contexts	Technology provides limited connections between sensemaking and real-world contexts	Technology effectively connects sensemaking to relevant real-world contexts	Technology creates immersive experiences that deeply connect sensemaking to real-world contexts
Learning Beyond Class	Technology doesn't support sensemaking beyond class time	Technology provides limited opportunities for sensemaking beyond class time	Technology effectively supports sensemaking beyond class time	Technology seamlessly extends sensemaking beyond class, fostering continuous inquiry
Authentic Tools	Technology use doesn't reflect authentic scientific tools and practices	Technology somewhat reflects authentic scientific tools and practices	Technology use clearly reflects authentic scientific tools and practices	Technology use mirrors professional scientific tools and practices, providing highly authentic experiences

Tips for Using This Rubric:

1. **Engagement:**
   Watch for how technology affects student motivation and focus—does it enhance their investment in sensemaking or create unnecessary distractions?

Notice how technology supports social-emotional aspects of learning through collaborative sensemaking and shared exploration of phenomena.

2. **Enhancement:**
   Look for ways technology reduces cognitive load and provides scaffolding—does it help students organize their thinking so they can focus on deeper sensemaking?
   Pay attention to how technology deepens understanding through various features and tools that support the sensemaking process.
3. **Extension:**
   Document how technology connects classroom sensemaking to real-world contexts and authentic scientific practices.
   Watch for evidence that technology enables meaningful sensemaking to extend beyond class time.
4. **Use This Rubric Formatively:** Use this rubric both formatively and summatively to guide improvements in how technology supports student sensemaking.
5. **Adapt the Rubric as Needed:** Adapt the criteria for your specific grade level, subject area, and technology tools while maintaining focus on meaningful sensemaking.

Remember, the goal isn't simply to use technology because it's available—it's to thoughtfully integrate digital tools in ways that engage students with phenomena, enhance their sensemaking processes, and extend their learning beyond the classroom walls. When technology truly supports students' efforts to figure out scientific phenomena and solve problems, it becomes an essential part of their sensemaking toolkit rather than just another layer of classroom complexity.

Rubric—Assessing Mathematical Sensemaking

Mathematical sensemaking goes far beyond getting the right answer—it's about how students understand, approach, and reason through mathematical situations. This rubric helps us look at the multiple dimensions of mathematical thinking, from how students initially tackle problems to how they communicate their understanding. Whether students are exploring patterns, making connections, or explaining their reasoning, we're looking for evidence that they're actively making sense of mathematical ideas rather than just following procedures.

Criteria	Beginning (1)	Developing (2)	Proficient (3)	Advanced (4)
Problem-Solving Strategies				
Understanding the Problem	Shows limited understanding of the problem; may misinterpret key information	Demonstrates basic understanding of the problem, but may miss some nuances	Clearly understands the problem, including key information and constraints	Demonstrates deep understanding of the problem, including subtle aspects and implications
Approach Selection	Uses limited or inappropriate strategies to approach the problem	Selects basic strategies that partially address the problem	Chooses appropriate and effective strategies to solve the problem	Selects sophisticated strategies, possibly combining multiple approaches creatively
Persistence and Flexibility	Gives up easily when faced with challenges; reluctant to try alternative approaches	Shows some persistence; may try an alternative approach if prompted	Persists through challenges and willingly tries alternative approaches	Demonstrates high persistence, flexibly adapting approaches and seeing setbacks as learning opportunities
Mathematical Representations				
Use of Representations	Uses limited or inappropriate mathematical representations	Uses basic representations that partially capture the problem	Effectively uses appropriate representations to model the problem	Creates and uses multiple, sophisticated representations to gain deeper insights
Connections Between Representations	Makes few connections between different representations	Makes some basic connections between representations	Clearly connects different representations, using them to support understanding	Fluently moves between representations, using their interplay to gain deeper insights
Reasoning and Justification				
Logical Reasoning	Shows limited logical reasoning in solution steps	Demonstrates basic logical reasoning in solution steps, but may have gaps	Applies clear, logical reasoning in solution steps throughout the solution process	Demonstrates sophisticated reasoning in solution steps, considering multiple logical pathways
Justification	Provides little or no justification for solution steps	Offers basic justifications for solution steps, but may lack details or clarity	Clearly justifies solution steps with mathematical reasoning	Provides comprehensive, insightful justifications for solution steps, possibly anticipating and addressing potential questions

Criteria	Beginning (1)	Developing (2)	Proficient (3)	Advanced (4)
		Connections and Generalizations		
Mathematical Connections	Makes few connections to prior mathematical knowledge	Makes some basic connections to prior mathematical knowledge	Effectively connects the problem to relevant, prior mathematical knowledge	Makes insightful connections across multiple areas of mathematical knowledge
Real-World Connections	Shows limited ability to connect mathematics to real-world contexts and situations	Makes basic connections between mathematics and real-world contexts and situations	Clearly connects and relates mathematics to relevant, real-world contexts and situations	Demonstrates deep understanding of how mathematics connects, models, and informs real-world contexts and situations
Pattern Recognition and Generalization	Struggles to identify patterns or make generalizations	Identifies simple patterns, but may struggle to make generalizations	Recognizes patterns and makes valid generalizations	Identifies subtle patterns and makes powerful generalizations that extend beyond the immediate problem
		Mathematical Communication		
Clarity of Explanation	Provides unclear or incomplete explanations	Provides basic explanations with some clarity	Provides thoughtful, clear, and complete explanations	Provides exceptionally clear, concise, and insightful explanations
Use of Mathematical Language	Uses limited or inappropriate mathematical language	Uses some appropriate mathematical language	Consistently uses appropriate mathematical language and notation	Uses precise and sophisticated mathematical language, enhancing the clarity of communication
Audience Awareness	Shows little awareness of audience in making explanations	Demonstrates some awareness of audience, but explanations may not always be accessible	Explains ideas in ways that are generally accessible to the intended audience	Skillfully adapts explanations to the audience, making complex ideas accessible without oversimplification

Tips for Using This Rubric

1. **Problem-Solving Strategies:** Watch how students understand, approach, and persist with unfamiliar problems—are they making sense of problems and persevering, or searching for familiar procedures?

2. **Mathematical Representations:** Notice how students create, use, and connect different representations (equations, graphs, diagrams) as tools for mathematical reasoning and sensemaking.
3. **Reasoning and Justification:** Listen for students' reasoning and justification—are they constructing viable arguments and critiquing others' reasoning with clear logic?
4. **Connections and Generalizations:** Pay attention to how students connect mathematical ideas to patterns and structures, looking for generalizations that can help them solve new problems.
5. **Mathematical Communication:** Look for evidence of clear, precise, and effective mathematical communication that helps others understand their thinking.
6. **Use this Rubric Formatively:** Use this rubric formatively throughout units to guide improvements in students' mathematical sensemaking—not just as an end-point evaluation.
7. **Adapt the Rubric as Needed:** Adapt criteria and examples for your specific grade level and mathematical topics while maintaining focus on mathematical sensemaking practices.

Remember, mathematical sensemaking isn't about following a straight path to an answer—it's about developing rich understanding of mathematical relationships and patterns through exploration, reasoning, and communication. This rubric helps us capture the many ways students demonstrate their developing mathematical understanding.

Rubric—Assessing Sensemaking in Engineering

Engineering sensemaking differs from scientific sensemaking in important ways—while science focuses on figuring out how and why phenomena occur, engineering involves using that scientific understanding to solve problems through systematic design. This rubric helps us look at how students engage in engineering sensemaking, from analyzing problems through testing solutions. It's not about building towers from marshmallows or following prescribed design steps—it's about how students apply scientific principles to generate knowledge about what works and why, using that understanding to inform each iteration of their design. Whether students are analyzing stakeholder needs, applying scientific concepts, or testing prototypes, we're looking for evidence that they're actively making sense of both the problem and solution spaces through systematic investigation and evidence-based design decisions.

Criteria	Beginning (1)	Developing (2)	Proficient (3)	Advanced (4)
		Problem Definition		
Identifying Needs	Struggles to identify the core need or problem	Identifies basic needs, but may miss important aspects	Clearly identifies and articulates the core needs or problems	Demonstrates insightful understanding of needs, including subtle or underlying issues
Constraints and Criteria	Recognizes few or no constraints; few or no criteria	Identifies some basic constraints and criteria	Clearly identifies relevant constraints and criteria	Comprehensively analyzes constraints and criteria, including potential conflicts
Stakeholder Consideration	Shows limited awareness of stakeholders	Identifies main stakeholders, but may not fully consider their perspectives	Considers perspectives of various stakeholders in problem definition	Demonstrates deep understanding of diverse stakeholder needs and how they inform the problem
		Research and Analysis		
Information Gathering	Collects limited or irrelevant information	Gathers basic, relevant information	Effectively collects and organizes relevant information from multiple sources	Comprehensively collects information, synthesizing information from diverse, high-quality sources
Data Analysis	Performs limited analysis of data	Conducts basic analysis of data, with some relevant insights	Effectively analyzes data, drawing clear conclusions	Performs sophisticated analysis of data, revealing deep insights and patterns
Application of Scientific Principles	Shows limited application of scientific principles to the problem	Applies some basic scientific principles to the problem	Effectively applies relevant scientific principles to inform the problem	Demonstrates advanced application of scientific principles to the problem, integrating multiple concepts
		Ideation and Concept Development		
Generating Ideas	Generates few ideas, often conventional	Generates several ideas with some creativity	Produces multiple, diverse ideas showing creativity	Demonstrates exceptional ideation, producing numerous innovative concepts

Criteria	Beginning (1)	Developing (2)	Proficient (3)	Advanced (4)
Evaluating Concepts	Evaluates ideas superficially	Provides basic evaluation of ideas against some criteria	Systematically evaluates ideas against established criteria	Conducts comprehensive evaluation, considering multiple perspectives and potential trade-offs
Refining Concepts	Shows limited refinement of ideas	Refines ideas in basic ways when prompted	Effectively refines and combines ideas to improve concepts	Demonstrates sophisticated concept refinement, showing insight and creativity
		Design and Modeling		
Creating Representations	Creates basic or incomplete representations of designs	Develops representations that partially capture design ideas	Creates clear, appropriate representations of design ideas	Produces sophisticated, detailed representations that enhance understanding of the design
Use of Modeling Tools	Shows limited ability to use modeling tools	Uses basic modeling tools with some effectiveness	Effectively uses appropriate modeling tools to develop and communicate designs	Demonstrates advanced use of modeling tools, enhancing design and communication
Predicting Performance	Makes limited or inaccurate predictions about design performance	Makes basic predictions about design performance	Effectively predicts design performance based on models and analysis	Makes sophisticated predictions, considering complex interactions and edge cases
		Testing and Iteration		
Developing Test Plans	Creates limited or inappropriate test plans	Develops basic test plans that address some key aspects	Creates comprehensive test plans that address key design criteria	Designs sophisticated test plans that rigorously evaluate the design against all criteria
Analyzing Test Results	Performs limited analysis of test results	Conducts basic analysis of test results, identifying some key findings	Effectively analyzes test results, drawing clear conclusions for design improvement	Demonstrates insightful analysis of test results, revealing subtle issues and opportunities

Criteria	Beginning (1)	Developing (2)	Proficient (3)	Advanced (4)
Iterative Improvement	Shows resistance to design changes	Makes basic design improvements based on test results	Effectively uses test results to make meaningful design improvements	Demonstrates sophisticated iterative processes, making insightful improvements across multiple iterations
Communication and Collaboration				
Explaining Design Decisions	Provides limited explanations for design decisions	Offers basic explanations for some design decisions	Clearly explains and justifies key design decisions	Provides comprehensive, insightful explanations for all design decisions
Technical Communication	Uses limited technical language in communications	Uses some technical language in communications but may lack precision	Effectively uses appropriate technical language in communications	Demonstrates sophisticated use of technical language in communications, enhancing clarity and precision
Teamwork and Collaboration	Struggles to work effectively in a team	Contributes to team efforts with some effectiveness	Works effectively as part of a team, contributing to shared goals	Demonstrates exceptional teamwork, enhancing team performance and fostering collaboration

Tips for Using This Rubric

1. **Problem Definition:** Watch for how students make sense of the problem space—do they consider needs, constraints, and stakeholders as interconnected aspects of the engineering challenge?
2. **Research and Analysis:** Notice how students gather and analyze information—are they using research and data to inform their understanding of the problem and potential solutions?
3. **Ideation and Concept Development:** Pay attention to students' ideation process—how do they apply scientific principles to generate and evaluate ideas?
4. **Design and Modeling:** Look for how students use various representations and models to make sense of their designs and predict performance based on scientific understanding.

5. **Testing and Iteration:** Document how students make sense of test results—are they using data to inform meaningful iterations rather than making random changes?
6. **Communication and Collaboration:** Notice how students explain their design decisions—do they connect their choices to evidence and engineering principles?
7. **Use This Rubric Formatively:** Use this rubric formatively throughout the engineering design process to guide improvements in students' sensemaking.
8. **Adapt the Rubric as Needed:** Adapt the criteria for your specific grade level and project context while maintaining focus on evidence-based engineering design and scientific understanding rather than simple building activities.

Remember, engineering sensemaking is about applying scientific understanding to solve problems through systematic investigation and design. This rubric helps us capture how students use evidence and scientific principles to inform their design decisions, not just their ability to build a fancy final product.

CHAPTER 5

Developing Teacher Capacity for Sensemaking

As a result of our exploration of sensemaking so far in this book, we have learned that sensemaking is not just another educational buzzword. It actually represents a fundamental shift in how we approach STEM education. For this shift to truly take root in our classrooms, we need to focus on where educational change begins, and that is with our teachers. In this chapter, I delve into the critical task of developing teacher capacity for sensemaking, both at the preservice level in our university teacher education programs and with inservice teachers already working in the field.

Let's start with our university methods courses, the area of study where many future teachers first grapple with the "how" of teaching. Traditional approaches to these courses often focus on delivering content and managing classrooms, but sensemaking demands something more. It requires a reimagining of what it means to teach science, technology, engineering, and mathematics. Having taught both Elementary and Secondary Science Methods courses for over a dozen semesters, I've seen firsthand the transformative power of practice-based approaches in preparing preservice teachers for sensemaking-centered STEM classrooms. In each of these courses, I move beyond simply telling future teachers about sensemaking. Instead, I immerse them in the experience itself. In other words, we educators can create environments in which preservice teachers can engage in sensemaking themselves, grappling with phenomena and problems just as their future students will do. This might involve analyzing real classroom videos, role-playing challenging classroom scenarios, or designing and critiquing sensemaking-focused lesson plans. The key is to make these experiences as authentic as possible, allowing preservice teachers to develop not just theoretical knowledge, but also practical wisdom about how sensemaking unfolds in real classrooms.

Recently, I co-created a course at Boston University called "Sensemaking in STEM" with colleagues Eve Manz (Math & Science Education) and

Alejandra Salinas (Math Education). The course focuses on many of the ideas in this book. To illustrate the similarities between the book and the course, I have listed the following key points that underscore the course objectives. Students will:

- Reflect on their ideas around sensemaking in STEM and how they have shifted.
- Develop and engage in a sensemaking conversation around STEM phenomena and ideas.
- Attend to and interpret other students' sensemaking of STEM-related topics and identify the questions, ideas, and interests they bring to the topic by applying asset-based and culturally expansive perspectives.
- Design a learning resource or experience that supports sensemaking about mathematics, science, and/or an engineering phenomenon or design problem.
- Explain why sensemaking in STEM is crucial to contributing to a socially just world.

Courses like this one are crucial in helping our future educators see the complexities of sensemaking. This includes developing the ability to effectively design and lead sensemaking learning experiences in their own classrooms. Interestingly, I am currently teaching this course for the second time during the Fall 2024 semester. While I have typically taught students majoring in STEM education or math education, I have also taught students majoring in computer science, communication, and advertising. Regardless of their major disciplines and primary interests, they all saw the value in learning more about how people make sense of complex things in their world. Our discussions have been rich and meaningful, demonstrating students' attempts to make sense of what the process of sensemaking involves across various contexts. In addition, their final projects have been inspiring—from one student's beautifully self-illustrated exploration of computer graphics that made complex algorithms accessible through storytelling, to another's practical guide helping high school math teachers foster authentic sensemaking discussions in their classrooms, to a thoughtfully designed children's book that used sustainable materials to teach young visitors about ocean conservation at the New England Aquarium. These projects demonstrate creative ways of making STEM concepts meaningful through multiple modes of expression while maintaining scientific accuracy. Students thoughtfully considered how different audiences might connect with and make sense of complex ideas, creating resources that bridge technical understanding with everyday experiences. Seeing my students develop such varied approaches to making STEM concepts accessible reinforces the importance of diverse teaching methods. As STEM teachers whose aim is to incorporate sensemaking in our instructional methods, it is vital that we address how we can better facilitate this kind of teaching and learning process for broader and more diverse groups of lifelong learners.

Now, what about the millions of teachers already positioned in K–12 classrooms across the nation? The idea of reaching inservice teachers with new approaches to teaching presents a particular challenge, especially given the understandable skepticism many experienced educators have developed toward "the next big thing" in education. These teachers, representing a wide range of instructional experience and years on the job, need more than just exposure to new ideas—they need compelling evidence, sustained support, and practical pathways for implementation. This is where thoughtfully designed and facilitated professional development comes into play. Effective professional development for sensemaking goes beyond one-off workshops or presentations. It requires sustained engagement, opportunities for practice and reflection, and, most importantly, a supportive community of peers. We need to create spaces in which teachers can experience sensemaking for themselves, reflect on their current practices, and collaboratively develop new approaches. This might involve long-term professional learning communities, coaching relationships, or action research projects through which teachers would investigate the impact of sensemaking implementation in their own classrooms. At the same time, it is important to acknowledge the fact that teachers do not work in isolation. For sensemaking to truly take root, we need to consider the entire ecosystem of education. This means engaging STEM coaches, administrators, and district-level leaders in the sensemaking journey. These educational leaders can play a crucial role in creating the conditions for sensemaking to flourish. They need to understand not just the "what" of sensemaking, but also the "why" and "how" in order to support the sensemaking approach to instruction. Thus, educational leaders need to be equipped to support teachers in this transition, to advocate for necessary resources, and to align policies and practices with sensemaking approaches.

In the sections that follow, I discuss specific strategies for leading professional development, supporting STEM coaches and administrators, and implementing sensemaking at the district level. I draw on research-based best practices and real-world examples to provide a road map for developing teacher capacity for sensemaking at all levels of the educational system. Ultimately, if we want our students to engage in deeply meaningful sensemaking in STEM disciplines, we need to ensure that our teachers are equipped, supported, and inspired to lead the way.

METHODS COURSES

I absolutely love teaching. I truly cherish the teacher–learner relationship that unfolds with each of my students. Early on, I share with them that I aim to get to know them so that I can better understand their own personal perspectives regarding what they hope to learn and how I can help them be successful in that process. In this section, my aim is not to describe everything about how to teach STEM methods courses. Rather, I aim to highlight how sensemaking can become a core component of the wonderful work that is already being done in methods courses across the country.

One of the best professional communities to which I belong is the National Elementary Science Teacher Educator Group (NESTeg). Members of NESTeg include elementary science methods faculty from across the country, representing both large and small universities located in both urban and rural areas. This group, led by Christina Schwarz (professor of Teacher Education at Michigan State University), began during the pandemic as a professional learning community dedicated to developing more knowledge about equitable practices and strategies for leading online science methods courses. Typically, between 10 and 35 faculty members and graduate students join the monthly virtual meetings, with their educational ranks ranging anywhere from doctoral students to full professors who have been in the field for decades. Participants consistently discuss a number of problems of practice as well as opportunities for growth. The topics of discussion span multiple areas impacting preservice teacher education, including:

- Enhancing science education in elementary schools
- Increasing enjoyment and engagement in science for teachers
- Fostering relationships in classrooms and communities
- Promoting social justice and equity in elementary science education
- Addressing environmental justice issues at elementary levels
- Developing pedagogical approaches, tools, and frameworks
- Solving challenges in elementary science methods teaching

Professional learning communities like NESTeg provide their members with key opportunities to carefully examine their own instructional practices and learn from their peers, many of whom have been doing amazing work within their own professional networks. Sensemaking often emerges as a discussion topic, especially targeting the need for our methods students to experience STEM instruction in a new way, a way that is very different from how many of these doctoral students and professors were taught in STEM courses during their years of schooling. Still, methods courses are limited in terms of time and the topics that can be covered. Consequently, including something new (i.e., the sensemaking approach) always means removing something else. For this reason, I next highlight the elements of the sensemaking approach that I have found to be crucial for methods students to experience. In turn, they need to know how to design and lead lessons grounded in this type of teaching and learning process.

SENSEMAKING REHEARSALS

Earlier in this book, I discussed and demonstrated the importance of mini-reasoning experiences that I consider key to the development of pre- and inservice teachers' understanding of how to implement the sensemaking approach in their current and future classrooms. Now I shift to highlighting a version of mini-reasoning experiences known as *rehearsals*. Rehearsals provide a unique and powerful learning environment for preservice teachers to develop their skills in facilitating sensemaking discussions in science classrooms. By approximating real classroom interactions in a controlled setting, rehearsals allow novice teachers to practice responding to student ideas and how to guide a classroom of students toward collective meaning-making around scientific phenomena. At the same time, the ability to pause and receive immediate feedback during rehearsals is particularly valuable for developing preservice teachers' capacity for in-the-moment decision-making. As Davis et al. (2017) found, teacher educators strategically use pauses to problem-solve, provide targeted feedback, and highlight key aspects of practice. This allows novice teachers to refine their instructional moves and try out alternative approaches in real time. The interactive nature of rehearsals helps preservice teachers learn to attend closely to the dynamics of student thinking and adjust their facilitation accordingly. Additionally, rehearsals create opportunities for preservice teachers to experience sensemaking from both the teacher and student perspectives. By taking on student roles during their classmates' rehearsals, teacher candidates gain insight into how learners engage in the sensemaking process. This dual experience enhances their ability to anticipate and respond productively to student ideas during discussions.

The customized nature of rehearsals allows teacher educators to tailor their guidance to each preservice teacher's specific needs and context. This personalized coaching helps novices develop their own style of facilitating sensemaking while adhering to core principles of the practice. At the same time, the collective aspect of rehearsals enables preservice teachers to learn vicariously by observing and reflecting on their peers' attempts at facilitating discussions. In this way, rehearsals provide a safe space for preservice teachers to take risks, make mistakes, and iteratively refine the practice of guiding sensemaking. The focused, deliberate practice opportunities afforded by rehearsals allows novices to develop fluency with key moves and principles for facilitating productive science discussions. This targeted preparation equips them to more effectively support student sensemaking once they lead real classrooms of students.

In my elementary methods course, we are on site for a whole school day at the William Monroe Trotter K–6 School in Dorchester, Massachusetts. Rather than simply learning about teaching science solely from research articles and academic texts, I strive to balance the theory we are learning with actual practice in the classroom—with real Boston Public Schools (BPS) students. In this context, I co-teach with Brenda Richardson, a veteran science teacher who has been working with Boston University students, allowing them to practice their skills in her classroom since 2006. There could, and probably should, be a whole book about how beneficial this kind of authentic teaching experience is for preservice teachers. For now, I want to focus on an example of how we move from idea (i.e., what sensemaking is), to the rehearsal experience, and then to my BU methods students actually leading a lesson in the Trotter School classroom.

BU students begin at the Trotter School by first being paired with a group of students, usually 1st- through 5th-graders, who have been selected to work with our methods class students over the entire semester of our course. For my methods students, our day usually involves a block of time to go over our readings, reflections, and course material before shifting gears to work with the students who come to meet us in the "BU Room" at Trotter (a dedicated space for this work). On our first day at Trotter, BU students interview the BPS students, asking them what they already know about the topic we will be covering during the semester. This question serves as a way to get to know the students, and also as a way to do a sort of formative assessment about their existing understanding of the topic.

In the particular year that the following example took place, we were asked to create lessons that supported student learning connected to the Full Option Science System (FOSS) curriculum from The Lawrence Hall of Science (University of California, Berkeley). To accomplish this, BU students did a deep dive into the Framework and the NGSS to determine the disciplinary core ideas in science for this topic and spend some time unpacking those ideas. As part of the unpacking process, BU students learned to identify the key components of each DCI and how they relate to the broader scientific concepts. They explored how these ideas build upon one another from elementary to middle school to high school, gaining insight into the vertical alignment of science education. This understanding is considered crucial for BU students' knowledge base in terms of preparing to work with their assigned Trotter School students, who may be at various stages in their scientific understanding. Once the BU students had established a better understanding of the topic and their students' current understandings, they could use this information to tailor their approaches, in this way ensuring that their lessons would build upon and extend the Trotter students' existing knowledge.

Before leading these lessons with Trotter students, we had rehearsals during which BU students led their lessons for their classmates and myself. In the time leading up to these rehearsals, I had already spent time with the students, helping them to refine their lessons and ensuring that the overall learning goals were sound. However, note that I often leave out elements of the lesson that I want BU students to uncover on their own, which I did in this example. At the start of the rehearsal process, each student presented an introduction to what they would be teaching. They also stated the key parts of their presentation for which they wanted to receive feedback. The feedback might be a guiding question, or a particular way they have organized an experiment, or even helping them to think about the focus of their discussion at the end. I also shared with the group that I—and the two students who were leading the lesson—could pause the lesson at any point to open up a quick discussion about particular pedagogical approaches or opportunities to attend to student sensemaking. Most novice elementary science teachers are often worried about knowing the science behind the lesson or the logistical challenges of the experiment, but these pauses are specifically intended to be in service of key moments of sensemaking and how to support different ways of student thinking.

In the rehearsal example I am referencing, one group of preservice teachers had initially focused their insect lesson on insect habitats. Their investigation was initially centered on having Trotter students draw what they saw in the habitats of both insects—pill bugs and blue death feigning beetles—and non-insects—earthworms and crayfish. While there is nothing fundamentally wrong with this focus, the core science idea was about understanding how organisms are well-adapted to their environments.

During this group's rehearsal, I paused the lesson to have a discussion about their lesson focus as we were all drawing what we *saw* in the different habitats. The point here is that as a key element of these rehearsals, the lesson needed to be as real as possible so that the BU students were not just acting out what leading a lesson could look like. Instead, they needed to lead the lesson with their focus on the learning goals. Because of this distinction, it has been my habit to always collect all the materials needed to do a full rehearsal of the lesson. In this case, it meant that I collected all of the organisms so that my BU students could actually see what it would be like leading a lesson using the organisms. The two students leading the lesson noticed that their peers were just drawing the habitats without considering anything about the organism itself. So I paused the lesson to hear what they (student peers) were thinking as they were drawing.

During this pause, I asked the students leading the lesson what their learning goals were and what evidence they would need to see from us (the simulated students) to convince them that they were on their way to accomplishing those learning goals. Together, we came to understand that in order to learn about how these organisms were particularly suited for each environment, we would need to shift the focus to both the organism *and* the environment. In other words, the characteristics and behaviors of each organism would be a better place to direct our focus for this lesson's learning goal.

The pauses we took during rehearsals were not just about the students who were leading the lesson. Their classmates were immersed in the lesson, too, and they went into these rehearsals knowing that they would need to fill out a feedback form detailing the following: (a) what they perceived the learning goals to be, and (b) what they actually learned by engaging in the lesson. Typically, when we pause during a rehearsal session, the students are often confused about something; or ready to share, but not given the opportunity to do so; or thinking about a misalignment with what they were experiencing; or needing to clarify the lesson's learning goals. The moments of pausing made the rehearsal process real for every member of the classroom community, as opposed to just benefiting the individuals leading the lesson. At the end of this particular lesson, I gave the whole class 5 to 7 minutes to silently fill out feedback forms where they would capture their in-the-moment feedback. Those leading the lesson would do the same. Following this process, I conducted a whole-group debrief, allowing the feedback to be shared aloud. This debrief concluded in our thinking about how we could shift the instructional moves in the actual lesson for the Trotter students. The Trotter lesson would be revised based on this feedback. In the image shown of the final lesson, as adapted from the rehearsal feedback, the BU students switched from a main focus on the habitat to the behaviors and characteristics of each organism that help it survive in its environment. In the image, you can see a real blue

Figure 5.1. A Blue Death Feigning Beetle on a Graphic Organizer

death feigning beetle doing the best acting job on this graphic organizer to convince everyone that it was indeed dead and to be left alone. Trotter students then took this information to the classroom rug for a discussion on how these specialized behaviors help the organisms survive in different environments.

To successfully facilitate these kinds of sensemaking rehearsals in your own methods classroom, following are some recommendations that can be gleaned from Davis et al.'s (2017) article, "Teaching the Practice of Leading Sense-Making Discussions in Science: Science Teacher Educators Using Rehearsals":

1. Use pauses strategically during rehearsals to provide feedback, problem-solve, and highlight important aspects of practice. The most common purposes for pauses are providing feedback and problem-solving.

2. Focus on key elements of facilitating science discussions during pauses, such as attending to student thinking, use of language, and how to execute specific instructional moves.
3. Tailor feedback and guidance to the individual needs of the rehearsing teacher candidate, while also considering how to make comments relevant to the whole group.
4. Be mindful of the number and frequency of pauses—avoid interrupting too often, but provide enough opportunities for feedback and reflection.
5. Connect rehearsal experiences to program goals, values, and conceptual tools introduced in methods courses.
6. Prepare candidates for rehearsals by modeling the practice, analyzing videos/transcripts, and having candidates plan their lessons in advance.
7. Frame rehearsals as low-stakes opportunities to take risks and try out teaching moves. Normalize the process of being paused and receiving feedback.
8. Have candidates rehearse practices they will soon enact with real students in their field placements.
9. Use tools and scaffolds during rehearsals, like summary tables or back-pocket questions, to support candidates' facilitation of discussions.
10. Include opportunities for candidates to reflect on and debrief the rehearsal experience afterward.
11. Consider having teacher educators play dual roles—facilitating the rehearsal while also acting as a student to surface important ideas/misconceptions.
12. Align rehearsals with the core practices and learning goals for science teaching emphasized in the methods course and program.

Rehearsals will feel awkward at first. The more you and your students practice, the better they will get and the more you will be able to identify the key moments when a pause could be most beneficial. When trying out the rehearsal process the first few times, it can be common to want to pause the lesson every few minutes—or not at all. This urge to pause will shift as you begin to more clearly perceive the key moments in the rehearsal, and you will learn to predict these moments as you co-construct the lessons with your methods students.

Throughout the semester working with the Trotter School, my methods students worked to create lessons that not only aligned with the school's curriculum, but that also supported development of the relevant DCIs, SEPs, and CCCs. They learned to design activities and discussions that would challenge younger students to think critically about scientific phenomena and encourage them to make connections between what they observed in those lessons and the underlying scientific principles. In essence, my methods students learned to prioritize sensemaking in ways that move this learning forward. This approach allowed my BU students to experience firsthand how theoretical understanding of educational standards translates into practical, engaging lessons. They learned to scaffold instruction, starting from the students' current level of understanding and gradually introducing more complex ideas. This process helped these future teachers develop the skills needed to support students in constructing a coherent and progressively sophisticated understanding of scientific concepts over time. Moreover, the ongoing nature of their work at the Trotter School enabled my BU students to witness and facilitate the growth in young students' scientific thinking over the course of the semester. Ultimately, my methods students had the opportunity to observe how well-designed, standards-aligned instruction can lead to meaningful learning outcomes, as such preparing them for their future roles as effective science educators. I firmly believe that all higher education students should have the opportunity to experience their methods courses embedded and taught in partner schools so as to experience this kind of learning directly.

PROFESSIONAL LEARNING FOR INSERVICE TEACHERS

While developing preservice teachers' capacity for sensemaking is crucial, we cannot overlook the millions of inservice teachers already shaping young minds in classrooms across the nation. The challenge of reaching an estimated 4 million teachers nationwide with new approaches to sensemaking is indeed daunting, but it's a challenge we must meet head-on if we want to see widespread adoption of these practices in STEM education.

While the aim of this chapter is not to summarize the elements of effective professional development (PD) for inservice teachers, it is crucial to note that the kinds of PD to which I am referring are not the types that we know to be ineffective. What I term ineffective approaches are often one-off workshops, generic seminars disconnected from classroom practice, or passive lecture-style presentations—all of which commonly fail to provide teachers with practical, actionable strategies they can immediately implement in their classrooms. In addition, ineffective approaches to PD typically lack the ongoing support of educational leaders, along with the follow-up necessary for meaningful, long-term changes in teaching practices. Recognizing that professional learning opportunities for inservice teachers are limited, the following discussion is nuanced and acknowledges that effective forms of PD for science, technology, engineering, and mathematics can take various forms. Moving forward, I first summarize what we know about effective professional learning before sharing recommendations for leading PD focused on sensemaking, as well as suggestions for STEM coaches and district administrators.

What We Know About Professional Development

Professional development (PD) in education has been a subject of significant investment and attention, yet its implementation and effectiveness have often fallen short of expectations. Despite substantial financial commitments at the national level, many approaches to PD have been criticized for being superficial, disconnected from teachers' daily classroom experiences, and failing to align with established principles of adult learning (Ball & Cohen, 1999; Ball & Forzani, 2011; Luft et al., 2015). This disconnect is particularly evident in science education, an area in which teachers typically dedicate only a small fraction of their time to professional development activities. A major study revealed that merely one-third of secondary science teachers engaged in more than 35 hours of PD over a 3-year period (Banilower et al., 2018). Meanwhile, the same study found that over 40% of elementary teachers did not participate in any science-specific PD during the same time frame.

This problem is amplified because the landscape of professional development in the United States has become characterized by its broad array of topics and lack of coherence. My incredibly knowledgeable and lovely doctoral adviser Suzanne Wilson (2013) aptly described PD as a "carnival of options," while Hill (2007) referred to it as a "hodgepodge of providers, formats, philosophies, and content" (p. 114). This fragmented approach has led to inconsistent quality and effectiveness across PD programs. The situation has been so dire that Sykes (1996) went so far as to declare the lack of quality PD as "the most serious unsolved problem for policy and practice in American education today" (p. 465), and this still rings true today. However, the importance of professional development cannot be overstated. As Guskey (2000) noted, "One constant finding in the research literature is that notable improvements in education almost never take place in the absence of professional development" (p. 4). The challenge lies in designing and implementing PD that is truly effective and impactful. Over time, scholars have reached a consensus on the essential features of effective and sustainable professional development. These features include (a) a targeted focus on specific subject-matter content; (b) engaging teachers in active learning that closely relates to their classroom realities; (c) ensuring coherence with teachers' prior knowledge and beliefs, as well as with reform documents and policy; (d) providing sufficient duration in terms of intensity and contact hours; and (e) fostering collective participation of a critical mass of teachers within a school or department (Borko, 2004; Desimone, 2009, 2011; Garet et al., 2001; Putnam & Borko, 2000; Roth et al., 2011; Wilson, 2013).

Building on this foundation for effective professional development practices, Darling-Hammond et al. (2017) expanded on the features listed above to provide additional specificity about effective PD. Their review identified seven key characteristics of effective professional development practices.

Seven Characteristics of Effective PD

1. Content-focused
2. Incorporates active learning utilizing adult learning theory
3. Supports collaboration, typically in job-embedded contexts
4. Uses models and modeling of effective practices
5. Provides coaching and expert support
6. Offers opportunities for feedback and reflection
7. Of sustained duration (Darling-Hammond et al., 2017)

In recent years, there has been a growing interest in teacher leadership as a means to enhance education and professional development (Wenner & Campbell, 2017). Studies have shown that teacher leaders can play a crucial role in supporting learning opportunities for their colleagues, helping to improve overall teacher quality and contributing to sustainable school improvement efforts (Hunzicker, 2017; Lumpkin, 2016). A review by Kennedy (2016) found that the most effective PD programs were those offered by individuals or groups who had (a) extensive experience working with teachers, (b) a deep understanding of the challenges teachers face, and (c) programs based on their personal experience and expertise. In the field of science education, such experienced individuals are considered "science teacher leaders." Whitworth et al. (2022) defined a science teacher leader as "A teacher of science, who influences others while developing their leadership identity, and who uses their social, cultural, and symbolic capital to advocate for science and promote student learning" (p. 257). This definition highlights the multifaceted role that science teacher leaders can play in enhancing professional development and, by extension, the quality of science education.

Effective professional development practices across other STEM disciplines have certain unique characteristics of their own. In mathematics, effective PD emphasizes deep understanding of mathematical concepts and relationships, requiring facilitators to possess both specialized content knowledge and pedagogical content knowledge (PCK) (Borko et al., 2014; Jacobs et al., 2017). Active learning that involves engaging teachers in problem-solving and mathematical thinking is crucial for professional development, along with sustained, long-term PD and collaborative group work (Borko et al., 2014; Darling-Hammond et al., 2017; Desimone, 2009). Technology-focused PD prioritizes hands-on experience with tools and software, emphasizing integration into subject-specific teaching rather than mere tool use (Koehler & Mishra, 2009; Lawless & Pellegrino, 2007). Key components of effective technology PD are (a) continuous technical and pedagogical support, (b) differentiated instruction accounting for varying technological proficiencies, and (c) context-specific approaches considering available resources (Ertmer & Ottenbreit-Leftwich, 2010; Kopcha, 2012; Mouza, 2009). Engineering PD often adopts an interdisciplinary approach, integrating science and mathematics concepts (Daugherty & Custer, 2012). Design-based learning that incorporates engineering processes, real-world applications, and collaboration with industry professionals are features of engineering PD that enhance teacher engagement and provide valuable insights (Capobianco & Rupp, 2014; Guzey & Moore, 2015; Nadelson et al., 2013). Effective engineering PD also addresses common misconceptions about the field (Hsu et al., 2011).

As we continue to grapple with the challenges of providing effective professional development across STEM disciplines, it is clear that a more nuanced, context-sensitive approach is needed. By leveraging the expertise of experienced teacher leaders, grounding PD in authentic classroom experiences, and remaining responsive to the day-to-day realities of teaching, we can work toward creating professional development programs that truly support teachers and, ultimately, enhance student learning. In particular, the seven characteristics of effective PD identified by Darling-Hammond et al. (2017) continue to provide a robust framework for designing and implementing such programs, thus ensuring that STEM PD programs are content-focused, active, collaborative, model-based, supported by coaching, reflective in nature, and sustained over time. By adhering to these principles and remaining attentive to the specific needs and contexts of both preservice and inservice teachers, we can hope to overcome the historical shortcomings of professional development and create truly transformative learning experiences for current and future educators.

LEADING PROFESSIONAL DEVELOPMENT PROGRAMS FOCUSED ON SENSEMAKING FOR INSERVICE TEACHERS

The ability to effectively lead professional development programs for inservice teachers requires a delicate balance of respecting the existing expertise of educators while also introducing new concepts and practices. When it comes to sensemaking, many teachers could find that it challenges their traditional notions of what science teaching looks like. Effective professional development in this area should start by acknowledging this potential discomfort and framing sensemaking not as a replacement for existing practices, but as an enhancement that can deepen student learning and engagement in science education.

A key aspect of planning professional development for sensemaking is to provide teachers with opportunities to engage in sensemaking activities themselves. Just as we immerse preservice teachers in sensemaking experiences, inservice teachers also benefit from grappling with phenomena and working through the sensemaking process as learners. This firsthand experience can be eye-opening, helping experienced teachers understand the cognitive and social processes involved in sensemaking and how to better support their own students through this process. This kind of immersive learning is front and center in the Next Generation Science eXemplar (NGSX) system of professional learning. Each of our pathways of learning center on figuring out phenomena that enable teachers to engage as adult learners challenged to authentically figure out science ideas. These experiences are different from just pretending that participants do not know the content. Instead, this kind of training highlights active engagement with specific ideas in science that are often more complex than one might first anticipate (e.g., kinetic molecular theory, climate change, etc.). The NGSX approach exemplifies this principle by creating knowledge-building communities in which teachers participate through collaborative problem-solving activities and interactive dialogue around authentic, scientific phenomena. By positioning teachers as active learners and epistemic agents, this approach allows them to experience the same type of learning environment they are expected to create for their students. Professional development should also focus on helping teachers develop specific skills and strategies for facilitating sensemaking. This includes learning about talk moves and other discussion techniques that encourage students to articulate their reasoning, build on one another's ideas, and engage in collective knowledge-building activities. Providing teachers with concrete tools and opportunities to practice these skills in a supportive environment is crucial for building their confidence and capacity to implement sensemaking in their classrooms.

While NGSX professional learning pathways are composed of multi-day, deeply immersive experiences, there still exists an additional need for shorter introductions to sensemaking for educators (i.e., teachers, administrators, coaches) so they can experience it firsthand. These introductions often take the form of conference sessions at statewide or national STEM education conferences and serve as a way to demonstrate that this type of learning is indeed different from the norm and requires further PD opportunities like NGSX. For these types of introductions, one strategy that I have found to be most effective is the use of spinning tubes. These are simply small segments of PVC pipe (1.5–4″ long) that have an "X" and an "O" drawn on each opposing end. When sharing this activity, I ask groups of educators to gather around each table and simply make note of what they observe when I spin the tubes. To do this, you put your index finger over either the X end or the O end and push down, causing the tube to rapidly spin around and around, staying mostly in the same place on the table while spinning. It can take a few minutes to get this spinning process down, so make sure you practice a little beforehand! As you do this, groups of teachers will notice that if you push on the X end, you see the X as the tube is spinning—but you do not see the O! If you push on the O end of the tube, you will see that letter and not the X. This is perplexing because observers will notice that the other character is in the same position, just on the other side of the tube. Because it is also facing upward, you innately know that you should see it as well.

As you circulate around to each group to get them started, have each group go public with what they are observing. Then you want each group to clearly articulate what the phenomenon is that they are observing. This will feel awkward at first, especially for educators who are new to this type of sensemaking. They may start by saying, "Something weird is happening" or, "Ahh, this makes sense because of inertia, or momentum, or centripetal forces." Push more on having each group clearly articulate what the phenomenon is, while also explicitly naming what is confusing to them, in this sense encouraging them to move beyond simply trying to name fancy science vocabulary words that are often used as substitutes for actual conceptual meaning. Ideally, you want each group to work together to say something like, "Why can we only see the letter that we push on when it spins and not the other letter that we know is on the other side of the tube?" Once they have come up with a version of this sort of question, I bring the groups together to share their thoughts as an entire learning community. This is important because as each group shares, we expect everyone in the room to hear a variety of different ideas. This exchange of ideas and perspectives teaches us that even slight differences in how we state the phenomenon are important. Ultimately, we need to make sure that as a learning community, we are all talking about the same thing. In other words, as a result of the whole-group discourse experience, we need to achieve consensus and state the phenomenon, which is especially important because we need to ensure that all learners are making sense of the same thing before we dive back into sensemaking. For example, with the spinning tube, some groups notice that when you spin it fast, the tube does not spin flat on the table but actually bounces up (vertically) as it rotates. This is something that might help us figure out what is going on. However, this is not the phenomenon we are all trying to figure out. Having each group clearly state the phenomenon helps to ensure that we are all having a shared experience and that our next steps will make sense to every learner in the room. This process can sometimes feel redundant, but do not skip it—in PD or in classrooms—because it can easily lead groups astray, and it will take more time to refocus.

Once it is clear that everyone understands what we are making sense of, I ask PD participants, "Okay, so what next? We all know what we are trying to figure out. How should we get started?" I invite all participants to either think quietly for a moment or chat with an elbow partner. Then I ask them to share their thoughts and ideas with the whole group. As they share, I write down their ideas on the board or on a piece of chart paper to keep track of what has been said. Based on the numerous times I have demonstrated the spinning tubes activity, I have found that participants normally share some basic ideas and/or recommendations, such as (a) spend more time investigating the phenomenon, (b) use slow-motion mode on our camera phones to capture the movement of the spinning tube, and (c) spin the tubes on different surfaces to determine if the surface is a possible factor. Finally, there is usually one person in every group that says, "We should Google it!" Before moving on to investigating this phenomenon further, I stop to point out to the group that our list spans a number of the science and engineering practices. Investigating further and spinning on different surfaces represent additional explorations that can help us ask more questions and plan and carry out follow-up investigations. Recording slow-motion video is a form of collecting data that we can further analyze. Even Googling it—though especially unrewarding—is NGSS practice 8, "Obtaining, Evaluating, and Communicating Information."

Spinning Tubes

Strategies for Acquiring the Item:

- The best way to make these spinning tubes is to purchase a PVC pipe cutter, which looks like a pair of pliers/scissors, and then buy 1/2″×10′ or 12′ PVC (SCH 40 will be written on the pipe). Cut them into small tubes varying in size. Then use a permanent marker to write an "X" and an "O" on opposite ends of the tube. Depending on how many times you use the tubes, even the permanent marker will wear off from use, and you will eventually need to redraw these letters again.

How I Introduce the Spinning Tubes:

- Included in the surrounding text.

The Pedagogical Goals:

- The main goal with the spinning tube activity is to focus in on the following:
  - » CCC—Cause & Effect, Patterns, Structure & Function, Energy & Matter
  - » SEPs—Asking Questions, Planning and Carrying Out Investigations, Constructing Explanations, Designing and Using Models
  - » DCI—Forces and Interactions
- As a mini-reasoning experience, the spinning tubes activity is incredible because of its simplicity. It does not require much setup or any specialized materials and can easily be brought to conference sessions that require travel because these tubes barely take up space in your bag. Anyone can easily see the phenomenon, and it is very repeatable, yet it is much more complicated to explain than people think. This activity serves as a great introduction to three-dimensional learning because it shows the relationship between phenomena and the three dimensions that you need to begin making sense of.

Tips and Tricks for Your Success:

- When you push down on either end of the tube, you want your finger to snap to the table in a way that causes the tube to spin both along the long axis and along the middle. When you press on the X end, the tube will spin backward, rotating away from you while the tube itself is rolling forward. Think of it this way: If you were to only flick the tube with your finger, it would spin away. However, in this instance, you are pushing down on the tube, causing two motions simultaneously (spinning and rolling), but in opposing directions.

Depending on how much time you have with participants in a planned PD session, you can customize the spinning tubes activity as an introduction to sensemaking to meet your needs. For example, if a particular PD session is meant to be just a short introduction (as part of a multisession training) and you will be working with this group again, you can tailor your approach. When this is the case for me, I often give participants the tubes to take home and make sense of on their own before meeting for the next session. If the group is familiar with scientific modeling, I will ask them to create models of their own to explain this phenomenon, and we will use those in our next session. If I am not working with the same group more than once (e.g., a conference session or a professional series with different presenters), I will sometimes break my own rule and give the group the explanation. But I do not stand in the front of the room and explain it, fully robbing them of their *aha* moment. Instead, I pull up a slide that appears to be completely blank except for the title "Spinning Tube Explanation." In fact, the slide is not blank; rather, it contains full explanations—at both middle and high school levels (and beyond)—but all of the text is colored white against a white background. I do this to highlight the fact that answers exist, especially now, with better and better versions of generative AI that can quickly give you an answer to seemingly any question. Still, the journey is often more fun and rewarding than the destination. And this is especially true when it comes to learning. Reading answers online does not help us build knowledge. We learn when we struggle, work together, and make personal connections to what we currently know as we try to make sense of our world over time. When we do this, we are also building our own identities as sensemakers and building resilience in knowing that we can figure out many of the complex phenomena and problems we encounter. This sense of human resilience is especially important, as we cannot always count on the explanations we find in various places online. To test this proposition, I have asked multiple AI chatbots for an explanation of the spinning tube, and they all left out major components of the explanation. As I like to say, we can Google facts and definitions, but we cannot Google knowledge! Sensemaking is our pathway to building knowledge.

INFORMATION FOR STEM COACHES AND ADMINISTRATORS

When it comes to sensemaking practices in schools, STEM coaches and administrators play a key role in providing the resources (i.e., time and material) for this kind of professional learning to occur. But here's the thing:

these educational leaders need their own opportunities to experience sensemaking, too. They need to dive deep into the world of sensemaking in order to understand what sensemaking looks and feels like so that they can be the pillars of support that teachers need. For STEM coaches, this might involve mastering the art of planning and facilitating sensemaking-focused, lesson study cycles with teachers. Through collaborative planning, observation, and debriefing of lessons centered on student sensemaking, coaches can guide teachers in refining their practice and cultivating a shared vision of effective sensemaking within their schools' contexts. But let's not forget our administrators—they're not just there to dot the i's and cross the t's. They need to get in on this sensemaking action, too. Their professional learning should also be designed as a deep dive into the following: (a) spotting and cheering on sensemaking in classrooms, (b) aligning school policies and resources to give sensemaking the spotlight it deserves, and (c) becoming the ultimate sensemaking cheerleaders to parents and the wider community. In NGSX, we have a pathway—"The PLANS Principal Pathway: Principals Learn About, Network, and Support 3-Dimensional Science Learning"—that is specifically aimed at strengthening the sensemaking knowledge of school administrators, district instructional leaders, coaches, and curriculum supervisors interested in acquiring strategies and updated knowledge about the vision of the Framework and NGSS in order to support teachers and students. This pathway recognizes that transformative change in science education doesn't happen in isolation. It requires a concerted effort from leaders at all levels, working in harmony to create an ecosystem where this new approach to science education can thrive. By participating in this experience, educational leaders will be better prepared to position themselves and, by extension, their teachers and students at the forefront of exciting developments in science education.

For educational and administrative leaders to truly get excited about this new way of teaching and learning in STEM, they need time to play, too. It is crucial that educational leaders not only carve out some sacred time for teachers to embark on their own sensemaking quests, they should do the same for themselves. Along the way, coaches and administrators should also be channeling Jane Goodall's skills of observation and venturing into classrooms—not with clipboards and stern looks, but with curious eyes and open minds. When they witness the sensemaking happening across different grade levels (without the pressure of evaluation looming), coaches and administrators often become the biggest cheerleaders for this kind of teaching and learning process in K–12 schools. Ultimately, it's about equipping our school-based leaders with the know-how to be the ultimate sensemaking supporters. After all, you can't lead others down a path you haven't explored yourself. So let's champion these leaders as they embark on their own sensemaking journeys, paving the way for a school culture where curiosity reigns and learning never stops.

SUPPORTING SENSEMAKING AT THE DISTRICT LEVEL

At the district level, supporting sensemaking requires a systems approach. This might involve the following initiatives: (a) revising curriculum frameworks to explicitly include opportunities for sensemaking, (b) adjusting assessment practices to value the process of sensemaking as much as the final product, and (c) aligning professional learning opportunities across the district to support a coherent vision of sensemaking in STEM education. Districts should be careful in selecting STEM curricula to ensure that it has been reviewed and found to be of high quality. (Note that STEM curricula take different forms across the disciplines of science, technology, engineering, and math.) Districts should also play a key role in fostering teacher leadership around sensemaking. By identifying and supporting teacher leaders who are successfully implementing sensemaking practices, districts can create a network of in-house experts who can provide peer support and mentoring. These teacher leaders can facilitate professional learning communities, lead workshops, and serve as a resource for colleagues who are working to incorporate more sensemaking into their instructional practices.

FINAL WORDS: BUILDING CAPACITY FOR TEACHER SENSEMAKING

As we conclude this exploration of developing teacher capacity for sensemaking, it is clear that this journey is both challenging and essential. The path to widespread adoption of sensemaking practices in STEM education is not a straightforward one, but it's a journey well worth undertaking. We've seen that effective professional development for sensemaking goes beyond traditional workshops or seminars. It requires immersive experiences, sustained engagement, and a supportive community of practice. For preservice teachers, methods courses serve as a critical foundation by providing opportunities to experience sensemaking firsthand and developing the skills needed to facilitate its use in their future classrooms. In this context, rehearsals and embedded experiences in partner schools offer invaluable practice opportunities in a low-stakes environment. For inservice teachers, professional development must be carefully crafted to respect their existing expertise while also introducing new practices. Experiences like the spinning tube activity can serve as powerful entry points, enabling teachers to engage in sensemaking themselves and reflect on its implications for their teaching philosophies and practices. STEM coaches and administrators play a crucial role in this ecosystem of change. By participating in their own sensemaking experiences and learning to recognize and support these practices in classrooms, they become essential advocates for this approach. At the district level, a systems approach is necessary—aligning curriculum, assessment, and professional learning opportunities—to support a coherent vision of sensemaking in STEM education.

Throughout this process, we must remember that change takes time. Building capacity for sensemaking is not about quick fixes or one-size-fits-all solutions. It's about creating a culture of continuous learning and improvement, that is, a culture in which teachers, coaches, and administrators are all engaged in the ongoing process of making sense of their practices and the world around them. As we move forward, let's embrace the complexity and richness of sensemaking. Let's create educational spaces in which teachers can take risks, grapple with uncertainty, and build their identities as sensemakers. It is through this process that we can truly transform STEM education—fostering classroom environments that encourage curiosity, where students and teachers alike engage in the joyful struggle of figuring things out, and where we cultivate knowledge and the lifelong skills of sensemaking that will serve our students well beyond their school years.

Conclusion

A New Vision for STEM Education: Sensemaking as a Lifelong Pursuit

Now that we have reached the conclusion of our exploration of sensemaking in K–12 STEM education, it is important to reflect upon the profound implications of this pedagogical approach. Sensemaking is far more than a teaching strategy—it represents a transformative shift in how we view science education and its role in preparing future generations of students for their roles as contributing members of society.

In this final chapter, I call for a collective commitment to embracing sensemaking as a foundational philosophy in K–12 education. Rather than merely reflecting on the ideas discussed throughout this book, I urge educators, administrators, and policymakers to actively support teaching practices that cultivate a generation of informed, adaptable, and intellectually curious learners. As these student sensemakers move beyond their years of schooling and enter the world of work, they will become sensemaking adults who can navigate complexity, engage in thoughtful problem-solving, and contribute to building a more resilient, innovative future for themselves, their families, their communities, and generations to come.

Imagine a classroom alive with curiosity. Students gather around a puzzling device, eyes bright with excitement. They aren't passively memorizing facts. On the contrary, they're hypothesizing, debating, and experimenting. They are actively engaged in or "doing" the process of learning about phenomena through exploration, cooperation, and interactive communication. This is what sensemaking looks like in action, and it is reshaping science education. Why does it matter? Because in a world overflowing with both information and misinformation, the ability to think critically and navigate complex issues is more than just an educational goal—it's a societal necessity. Contemporary issues like climate change, global pandemics, and artificial intelligence aren't just problems for scientists. In actuality, they are global challenges that require every individual to engage in thoughtful and responsible sensemaking and problem-solving activities. Therefore, sensemaking is not solely about producing more scientists (although that's a great outcome). More broadly, it's about cultivating scientific literacy for all. It is a teaching and learning skill (and mindset) that empowers students from every background to question, investigate, and reason while building various other skills that will benefit them in any field—from physics to public policy. Whether or not they remember the periodic table years from now, students will undoubtedly need to evaluate conflicting information, make evidence-based decisions, and understand how systems work. That's the essence of STEM sensemaking—a skill learned and practiced in K–12 classrooms that evolves and transcends academic disciplines to incorporate aspects of living and working that will serve individuals throughout their lives.

By embracing the sensemaking approach to teaching and learning, we educators are doing more than updating how we teach science. We are, in fact, fostering a generation of critical thinkers, problem-solvers, and lifelong learners who will be prepared not just for exams, but for the complexities of a rapidly evolving world. So, let's set aside outdated models of science education and turn our classrooms into spaces of exploration and discovery. When we empower students to make sense of the world around them, we aren't just shaping better scientists—we are shaping engaged, informed citizens ready to meet the challenges of the future.

The sensemaking revolution has arrived. Let's lead the way!

References

Ball, D. L., & Cohen, D. K. (1999). Developing practice, developing practitioners: Toward a practice-based theory of professional education. In G. Sykes & L. Darling-Hammond (Eds.), *Teaching as the learning profession: Handbook of policy and practice* (pp. 3–32). Jossey-Bass.

Ball, D. L., & Forzani, F. M. (2011). Building a common core for learning to teach: And connecting professional learning to practice. *American Educator, 35*(2), 17–21, 38–39.

Banilower, E. R., Smith, P. S., Malzahn, K. A., Plumley, C. L., Gordon, E. M., & Hayes, M. L. (2018). *Report of the 2018 National Survey of Science and Mathematics Education*. Horizon Research, Inc.

Borko, H. (2004). Professional development and teacher learning: Mapping the terrain. *Educational Researcher, 33*(8), 3–15.

Borko, H., Koellner, K., & Jacobs, J. (2014). Examining novice teacher leaders' facilitation of mathematics professional development. *Journal of Mathematical Behavior, 33*, 149–167.

Braaten, M., & Windschitl, M. (2011). Working toward a stronger conceptualization of scientific explanation for science education. *Science Education, 95*(4), 639–669. https://doi.org/10.1002/sce.20449

Campbell, T., Schwarz, C., & Windschitl, M. (2016). What we call misconceptions may be necessary stepping-stones toward making sense of the world. *The Science Teacher, 83*(3), 69.

Capobianco, B. M., & Rupp, M. (2014). STEM teachers planned and enacted attempts at implementing engineering design-based instruction. *School Science and Mathematics, 114*(6), 258–270.

Carson, R., & Lee, K. (1956). *The sense of wonder*. HarperCollins Publishers.

Cioè-Peña, M., & Snell, T. (2015). Translanguaging for social justice. *Theory, Research, and Action in Urban Education, 4*(1), 1–5.

Common Core State Standards Initiative. (2010). Common Core state standards for mathematics. http://www.corestandards.org/assets/CCSSI_Math%20Standards.pdf

Darling-Hammond, L., Hyler, M. E., & Gardner, M. (2017). *Effective teacher professional development*. Learning Policy Institute.

Daugherty, J. L., & Custer, R. L. (2012). Secondary level engineering professional development: Content, pedagogy, and challenges. *International Journal of Technology and Design Education, 22*(1), 51–64.

Davis, E. A., Kloser, M., Wells, A., Windschitl, M., Carlson, J., & Marino, J.-C. (2017). Teaching the practice of leading sense-making discussions in science: Science teacher educators using rehearsals. *Journal of Science Teacher Education, 28*(3), 275–293. https://doi.org/10.1080/1046560X.2017.1302729

Desimone, L. M. (2009). Improving impact studies of teachers' professional development: Toward better conceptualizations and measures. *Educational Researcher, 38*(3), 181–199.

Desimone, L. M. (2011). A primer on effective professional development. *Phi Delta Kappan, 92*(6), 68–71.

Ertmer, P. A., & Ottenbreit-Leftwich, A. T. (2010). Teacher technology change: How knowledge, confidence, beliefs, and culture intersect. *Journal of Research on Technology in Education, 42*(3), 255–284.

Ford, M. J. (2012). A dialogic account of sense-making in scientific argumentation and reasoning. *Cognition and Instruction, 30*(3), 207–245. https://doi.org/10.1080/07370008.2012.689383

Frazier, B. N., Gelman, S. A., & Wellman, H. M. (2016). Young children prefer and remember satisfying explanations. *Journal of Cognition and Development, 17*(5), 718–736.

Garet, M. S., Porter, A. C., Desimone, L., Birman, B. F., & Yoon, K. S. (2001). What makes professional development effective? Results from a national sample of teachers. *American Educational Research Journal, 38*(4), 915–945.

Grapin, S. E. (2024). The complex terrain of equity for multilingual learners in K–12 education. *Educational Researcher, 53*(3), 167–174.

Grapin, S. E., Suárez, E., & González-Howard, M. (2022). Developing a science teaching workforce prepared to teach multilingual learners. *Next Gen Navigator*. NSTA Press.

Guskey, T. R. (2000). *Evaluating professional development*. Corwin Press.

Guzey, S. S., & Moore, T. J. (2015, June). Assessment of curricular materials for integrated STEM education (RTP, Strand 4). In *Proceedings of the 2015 ASEE Annual Conference & Exposition* (Paper No. 12590). American Society for Engineering Education.

Hanuscin, D. L., Cisterna, D., & Lipsitz, K. (2018). Elementary teachers' pedagogical content knowledge for teaching structure and properties of matter. *Journal of Science Teacher Education, 29*(8), 665–692.

Hill, H. C. (2007). Learning in the teaching workforce. *The Future of Children, 17*(1), 111–127.

Hsu, M. C., Purzer, S., & Cardella, M. E. (2011). Elementary teachers' views about teaching design, engineering, and technology. *Journal of Pre-College Engineering Education Research, 1*(2), 31–39.

Hunzicker, J. (2017). From teacher to teacher leader: A conceptual model. *International Journal of Teacher Leadership, 8*(2), 1–27.

Jacobs, J., Seago, N., & Koellner, K. (2017). Preparing facilitators to use and adapt mathematics professional development materials productively. *International Journal of STEM Education, 4*(1), 30.

Kennedy, M. M. (2016). How does professional development improve teaching? *Review of Educational Research, 86*(4), 945–980.

Koehler, M. J., & Mishra, P. (2009). What is technological pedagogical content knowledge? *Contemporary Issues in Technology and Teacher Education, 9*(1), 60–70.

Kolb, L. (2011). *Triple E framework.* University of Michigan, School of Education.

Kopcha, T. J. (2012). Teachers' perceptions of the barriers to technology integration and practices with technology under situated professional development. *Computers & Education, 59*(4), 1109–1121.

Lawless, K. A., & Pellegrino, J. W. (2007). Professional development in integrating technology into teaching and learning: Knowns, unknowns, and ways to pursue better questions and answers. *Review of Educational Research,* 77(4), 575–614.

Liquin, E. G., & Lombrozo, T. (2022). Motivated to learn: An account of explanatory satisfaction. *Cognitive Psychology, 132,* 101453.

Lombrozo T. (2006). The structure and function of explanations. *Trends in Cognitive Sciences, 10*(10), 464–470. https://doi.org/10.1016/j.tics.2006.08.004

Lombrozo, T. (2018, September 23). *Explanation: The good, the bad and the beautiful* [Lecture]. Lehigh University, Bethlehem, PA, United States.

Lombrozo, T., & Carey, S. (2006). Functional explanation and the function of explanation. *Cognition, 99*(2), 167–204.

Lombrozo, T., & Wilkenfeld, D. (2019). Mechanistic versus functional understanding. In S. R. Grimm (Ed.), *Varieties of understanding: New perspectives from philosophy, psychology, and theology* (pp. 209–230). Oxford University Press.

Lovett, M. C., Bridges, M. W., DiPietro, M., Ambrose, S. A., & Norman, M. K. (2023). *How learning works: Eight research-based principles for smart teaching.* John Wiley & Sons.

Luft, J. A., Dubois, S. L., Nixon, R. S., & Campbell, B. K. (2015). Supporting newly hired teachers of science: Attaining teacher professional standards. *Studies in Science Education,* 51(1), 1–48.

Lumpkin, A. (2016). Teachers as leaders: The pathway to shared leadership. *The Educational Forum, 80*(3), 208–220.

Minstrell, J. (1992). Facets of students' knowledge and relevant instruction. In R. Duit, F. Goldberg, & H. Niedderer (Eds.), *Research in physics learning: Theoretical issues and empirical studies* (pp. 110–128). Institut für die Pädagogik der Naturwissenschaften an der Universität Kiel.

Moon, J., Michaels, S., & Reiser, B. J. (2012, November 20). Science standards require a teacher-learning rethink. *Education Week.*

Mouza, C. (2009). Does research-based professional development make a difference? A longitudinal investigation of teacher learning in technology integration. *Teachers College Record, 111*(5), 1195–1241.

Nadelson, L. S., Callahan, J., Pyke, P., Hay, A., Dance, M., & Pfiester, J. (2013). Teacher STEM perception and preparation: Inquiry-based STEM professional development for elementary teachers. *The Journal of Educational Research, 106*(2), 157–168.

National Academies of Sciences, Engineering, and Medicine. (2012). *A framework for K–12 science education: practices, crosscutting concepts, and core ideas.* The National Academies Press. https://doi.org/10.17226/13165

National Academies of Sciences, Engineering, and Medicine. (2021). *Call to action for science education: Building opportunity for the future.* National Academies Press.

National Research Council. (2013). *Next generation science standards: For states, by states.* National Academies Press.

National Science Teaching Association (NSTA). (n.d.). *Developing a science teaching workforce prepared to teach multilingual learners.* https://www.nsta.org/blog/developing-science-teaching-workforce-prepared-teach-multilingual-learners

NGSS Lead States. (2013). *Next Generation Science Standards: For states, by states.* The National Academies Press.

Odden, T.O.B., & Russ, R. S. (2019). Defining sensemaking: Bringing clarity to a fragmented theoretical construct. *Science Education, 103*(1), 187–205.

Putnam, R. T., & Borko, H. (2000). What do new views of knowledge and thinking have to say about research on teacher learning? *Educational Researcher, 29*(1), 4–15.

Reiser, B. J. (2013). What professional development strategies are needed for successful implementation of the Next Generation Science Standards. In *Invitational Research Symposium on Science Assessment* (pp. 1–22).

Roth, K. J., Garnier, H. E., Chen, C., Lemmens, M., Schwille, K., & Wickler, N.I.Z. (2011). Videobased lesson analysis: Effective science PD for teacher and student learning. *Journal of Research in Science Teaching, 48*(2), 117–148.

Steele, M. A., Halkin, S. L., Smallwood, P. D., McKenna, T. J., Mitsopoulos, K., & Beam, M. (2008). Cache protection strategies of a scatter-hoarding rodent: Do tree squirrels engage in behavioural deception? *Animal Behaviour, 75*(2), 705–714.

Sykes, G. (1996). Reform of and as professional development. *Phi Delta Kappan, 77*(7), 464.

Van Leeuwen, A., & Janssen, J. (2019). A systematic review of teacher guidance during collaborative learning in primary and secondary education. *Educational Research Review, 27*, 71–89.

Wenner, J. A., & Campbell, T. (2017). The theoretical and empirical basis of teacher leadership: A review of the literature. *Review of Educational Research, 87*(1), 134–171.

Whitworth, B. A., Wenner, J. A., & Tubin, D. (2022). Science teacher leadership: The current landscape and paths forward. *Handbook of Research on Science Teacher Education, 1*, 257–272.

Wilson, S. M. (2013). Professional development for science teachers. *Science, 340*(6130), 310–313.

Windschitl, M., Thompson, J., & Braaten, M. (2020). *Ambitious science teaching.* Harvard Education Press.

Index

About the Author

Thomas "TJ" McKenna is a clinical assistant professor of science education at Boston University's Wheelock College of Education & Human Development and the director of the Center for STEM Professional Learning at Scale. With over 20 years of expertise spanning scientific research, classroom teaching, informal education, and teacher preparation, Dr. McKenna is dedicated to bridging the gap between policy, practice, and research to enhance STEM education for all learners.

Dr. McKenna specializes in supporting preservice and inservice educators in delivering high-quality, phenomena-based science learning experiences aligned with the Next Generation Science Standards (NGSS). As the founder of Phenomena for NGSS (www.ngssphenomena.com), he has developed innovative resources to help teachers make science engaging and accessible for students.

A seasoned science communicator and educator, Dr. McKenna's on-air experience has honed his ability to distill complex scientific concepts into relatable and exciting opportunities for sensemaking. In his current role, he teaches STEM education courses, conducts research on scaling effective professional development, and leads initiatives to promote sensemaking in STEM education. His work continues to address the systemic challenges of equipping educators to inspire the next generation of STEM thinkers.

About the Author